U0931880

QUAKER
EST 1877

養生燕麥美味食譜

Nourishing Oatmeal Recipe

目 *contents* 錄

序言

香港營養師協會 ／ 12

全穀物特點面面觀 ／ 14
The Whole Truth About Whole Grains

全穀物的健康益處 ／ 15
The Health Benefits of Whole Grains

常見的全穀物及其製品 ／ 16
Common Types of Whole Grains and Whole-Grain Products

如何選購全穀物製品 ／ 17
How to Look for Whole-Grain Products

燕麥是一種全穀物 ／ 17
Oat Is a Whole Grain

燕麥的營養特點 ／ 18
Nutritional Characteristics of Oats

燕麥的健康益處 ／ 18
Goodness of Oats

燕麥 β- 葡聚糖醣的益生元效應 ／ 19
Prebiotic Effects of Oat β-glucan

將燕麥融入一日三餐 ／ 20
Fit Oats Into Your Meals

早餐系列 *Breakfast*

24 大蝦燕麥粥
Prawns Oat Congee

26 香菇雞肉燕麥粥
Shiitake and Chicken Oat Congee

28 芋香燕麥粥
Taro Oat Congee

30 番茄雞蛋燕麥粥
Tomato and Egg Oat Congee

32 淮山桂花燕麥粥
Chinese Yam and Osmanthus Oat Congee

34 時蔬蝦仁燕麥粥
Vegetable and Prawns Oat Congee

36 酒釀雞蛋燕麥粥
Oat Congee With Poached
Eggin Fermented Rice wine

38 南瓜燕麥蒸糕
Steamed Pumpkin Oat Cake

40 燕麥片糯米餅
Sticky Rice and Oat Cake

42 燕麥鱈魚餅
Codfish Oat Cake

44 太陽燕麥餅
Sweet Potato Oat Pancakes

46 芝士蟹柳燕麥可麗餅
Cheese Crabstick Oat Crepe

48 韭菜燕麥雞蛋餅
Chive and Egg Oat Pancake

50 五黑燕麥煎餅卷
Five Black Multi-Grain Pancake Roll

52 燕麥全麥饅頭
Steamed Whole-Wheat Oat Bun

54 燕麥洋蔥雞肉卷
Oat and Onion Chicken Roll

56 燕麥香蕉吐司
Banana Oat Toast

58 燕麥煎餃
Oat Fried Dumpling

60 燕麥素包
Veggie Oat Bun

62 五福燕麥燒賣
Wufu Oat Siu Mai

64 核桃燕麥米汁
Walnut Oatmeal Rice Milk

66 五紅補氣燕麥豆漿
Five Reds Nourishing Oat Soy Milk

68 五黑芝麻燕麥紫薯飲
Five Black, Sesame, and Purple Sweet Potato Oat Drink

70 奶香蘑菇配燕麥燴水波蛋
Creamy Mushroom With Oat and Poached Egg

72 燕麥厚蛋燒
Oatmeal Omelet

74 西蘭花燕麥烘蛋
Baked Egg With Broccoli and Oats

76 生椰拿鐵燕麥
Coconut Latte With Oats

78 燕麥藜麥鍋巴
Oat and Quinoa Crispy Rice

80 燕麥西蘭花薯餅
Oat and Broccoli Hash Browns

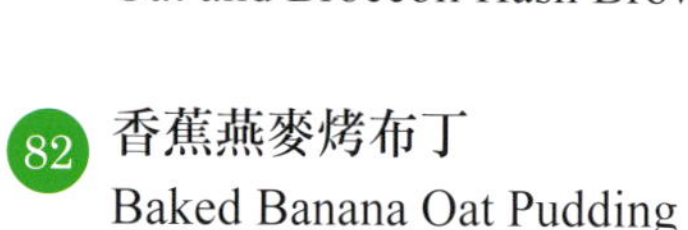

82 香蕉燕麥烤布丁
Baked Banana Oat Pudding

間餐系列 *Snacks*

86 栗蓉燕麥鬆餅
Chestnut Oat Muffins

88 紅薯燕麥甜甜圈
Sweet Potato Oat Donuts

90 抹茶燕麥餅乾
Matcha Oat Cookies

92 蔓越莓燕麥鬆餅
Cranberry Oat Scones

94 楊枝甘露燕麥撻
Mango, Pomelo, and Sago Oat Tart

96 豆乳桂花燕麥撻
Soya Milk and Osmanthus Oat Tart

98 抹茶燕麥糯米糍
Matcha Oatmeal Mochi

100 肉桂蘋果燕麥飲
Cinnamon Apple Oat Drink

102 桂花栗蓉燕麥飲
Osmanthus Chestnut Oat Drink

104 燕麥南瓜甘露
Pumpkin Oat Drink

午晚餐系列 *Main Dishes*

108 燕麥牛油果沙律
Avocado Oat Salad

110 燕麥香菌菇雞肉三文治
Mushroom and Chicken Oat Sandwich

112 燕麥蘆筍鮮蝦羹
Asparagus and Shrimp Oat Soup

114 豌豆泥小蘑菇沙律
Mashed Peas and Mushroom Salad

116 三文魚燕麥炒飯
Fried Rice With Oats and Salmon

118 豆腐燕麥茄汁燜飯
Tofu and Tomato Braised Rice With Oats

120 奶香菇燕麥燴飯
Creamy Mushroom Oat Risotto

122 燕麥福鼎肉片
Fuding Chicken and Oat Slices Soup

124 馬蹄燕麥蒸肉丸
Steamed Water Chestnut Oat Meatballs

126 燕麥乾煎三文魚
Pan-Seared Oat-Crusted Salmon

128 燕麥西蘭花三文魚忌廉湯
Salmon & Broccoli Oat Cream Soup

130 養生五色燕麥菜飯
Five-Colour Oat & Veggie Rice

132 黑胡椒雞粒炒燕麥飯
Black Pepper Chicken Fried Oat Rice

134 燕麥乾果餅卷
Oat & Dried Fruit Pancake Rolls

136 星洲黃金麥皮蝦
Singapore-Style Golden Oat Prawns

138 燕麥花雕肉餅蒸蟹
Steamed Crab with Huadiao Pork & Oats

序言

近年來，坊間對燕麥的健康價值出現不少誤解，甚至有人將其錯誤標籤為「不健康」的食物。然而，燕麥其實是極具營養價值的全穀物，可作為健康均衡飲食的一部分。

燕麥蘊含豐富的水溶性膳食纖維、維生素及礦物質。其中，大量科學研究證實，β-葡聚醣有助降低膽固醇、穩定血糖，並助維持心血管及腸道健康。相比精製碳水化合物（例如白米、白麵包），燕麥含有更多膳食纖維、植物化學物及其他微量營養素，適量食用能有效提升整體飲食質素。

市面上的燕麥產品種類繁多，營養價值亦存在差異。部分即食或調味燕麥產品加入了大量糖分、奶粉或其他添加劑，或會降低燕麥本身的健康益處。建議大家盡量選擇加工程度較低的燕麥，例如燕麥粒或原片燕麥片。這類產品成分相對天然，更能保留燕麥的營養價值。

另外，不少人以為燕麥味道單調、食法有限，其實燕麥極具變化。除了傳統的水煮燕麥粥，燕麥亦可以添加到許多不同的菜餚中，無論中式或西式料理都能輕鬆配搭。善用燕麥，不但能豐富每日膳食，還能滿足不同生活節奏和營養需要。

本食譜集由桂格團隊精心設計，收錄多款簡易、美味且營養均衡的燕麥食譜，幫助讀者跳出傳統框架，發掘燕麥的多元可能。香港營養師協會一直致力推廣以科學為本的營養知識，協助市民作出明智飲食選擇。我們誠意邀請你重新認識燕麥，並透過本食譜集，嘗試以創新方式將燕麥融入日常膳食，讓健康從每一餐開始。

香港營養師協會
2025 年

Disclaimer: The Hong Kong Dietitians Association (HKDA) has not received any honorarium from Quaker Oats Company and is not involved in any aspect of recipe development, vetting, or nutritional analysis. Any information provided above is for educational purposes only and does not constitute an endorsement of specific products or brands.

* 免責聲明：* 香港營養師協會（HKDA）並未從桂格燕麥公司收取任何酬金，也不參與任何食譜編寫、評論或營養分析工作。以上所提供的任何資訊僅供教育用途，並不構成對特定產品或品牌的認可。

關於燕麥
Oats Introduction

全穀物特點面面觀

在人類的飲食構成當中，全穀物的食用歷史最為悠久。穀物是膳食結構中重要的組成部分，是人們獲取能量、蛋白質、膳食纖維、維他命 B 雜與礦物質的主要來源。為了追求口味和美味，人們越來越喜歡選擇精製穀物及其加工食品作為日常主食，也就是說將穀物中原本存在的、富含膳食纖維、微量營養素與生物活性物質的麩皮和麥芽去除掉了，只攝入了主要含有澱粉及部分蛋白質的胚乳部分。但是，究發現，主食精細化加工與部分營養素攝入不足、腸道疾病、各種慢性非傳染性疾病的高發密切相關。[1] 越來越多的國家和機構，開始宣導更健康的穀物消費理念，鼓勵適量增加全穀物攝入，使膳食中全穀物與精製穀物的比例適當，又保留營養又不失美味，及時應對疾病譜的改變和營養素缺乏的「隱形饑餓」挑戰。

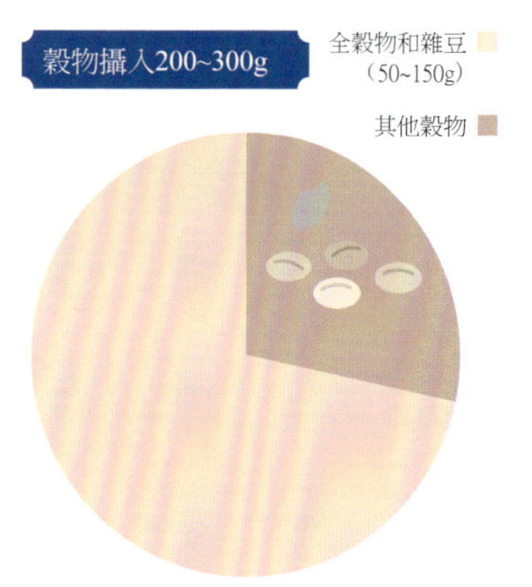

穀類為主是平衡膳食模式的重要特徵。中國傳統飲食習慣中作為主食的各種穀物，如稻類、麥類、玉米、高粱、黃米、蕎麥等，如果加工得當，均可作為全穀物的良好來源。中國居民膳食指南（2022）中指出健康的穀物攝入模式為：平均每天攝入穀類食物 200~300g，其中全穀物和雜豆類 50~150g。[2]但是，2020 年中國居民營養與慢性病狀況報告數據顯示，成年居民（>18 歲）以除大米、麵粉之外的粗糧和雜豆計算全穀物平均攝入量僅為 18.6g，僅佔穀物總量的 5%。[3]

全穀物

全穀物（whole grain）指經過清理但未經進一步加工，保留了完整穎果結構的穀物籽粒；或雖經碾磨、粉碎、擠壓等方式加工，但皮層、胚乳、胚芽的相對比例仍與完整穎果保持一致的穀物製品。換而言之，全穀物雖經不同程度的加工，但保留了完整穀粒的組織和結構，同時營養成分也得到了妥善的保存，特別是膳食纖維（如 β- 葡聚醣）、維他命（如維他命 B 雜，維他命 E）、礦物質（如鐵、鋅、鎂）、不飽和脂肪酸、植物甾醇、酚類化合物、蛋白質等。

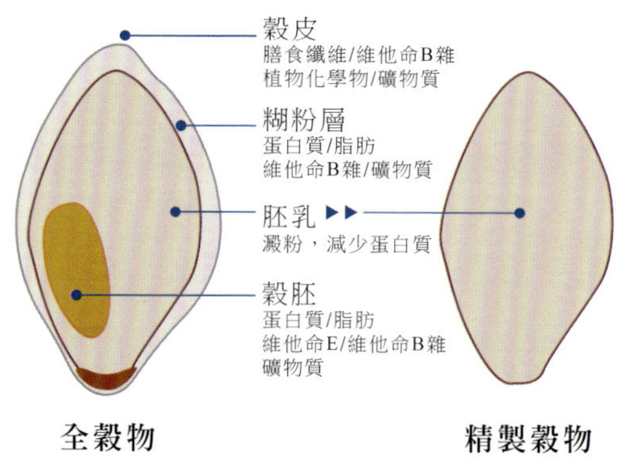

1 楊月欣譯：燕麥營養與技術[M]，北京：北京大學醫學出版社.2018-5。

2 中國營養學會，中國居民膳食指南（2022）[M]，北京:人民衛生出版社出版社，2022。

3 國家衛生健康委疾病預防控制局，中國居民營養與慢性病狀況報告[M]，北京：人民衛生出版社出版社，2020。

事實上，由於全穀物的健康效益得到了廣泛證實，包括中國在內的世界各地，均開始宣導更健康的穀物消費理念，鼓勵增加全穀物在穀物攝入總量中的比例，相對增加全穀物而減少精製穀物。

TIPS 全穀物和精製穀物的區別：出於對色澤、味道、口感等維度的考慮，現代穀物精加工技術除掉了穀物的麩皮和胚芽，使得穀物看上去更加白淨、晶亮，口感更細膩、軟糯。但這個過程將膳食纖維、微量營養素以及生物活性物質去除了，只留下了含有澱粉和蛋白質的胚乳，因此麩皮和胚芽中的所有其他重要營養物質全部流失殆盡。因此，在營養價值和促進健康層面，精製穀物只能甘拜下風。

全穀物的健康益處

隨著研究證據的積累，全穀物的各種健康效益正在被不斷發現及逐漸認可。目前研究較多的包括預防心血管疾病、2 型糖尿病、改善腸道健康、體重控制和免疫調節等。[6]

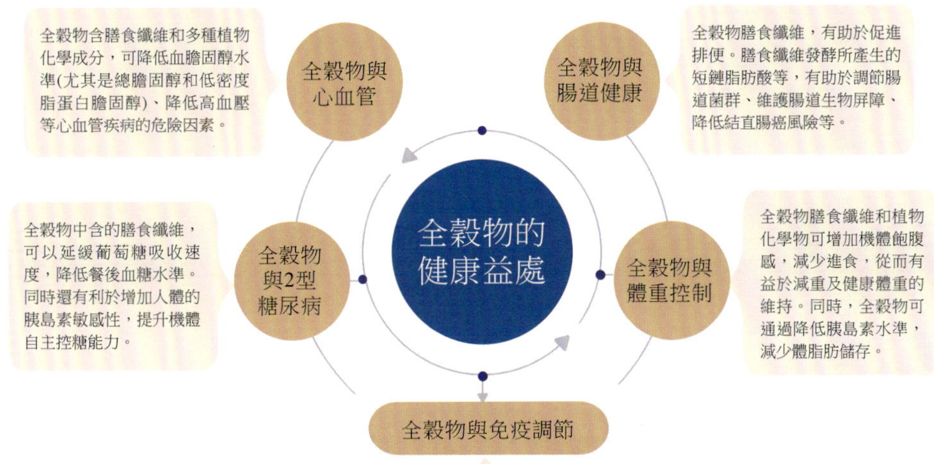

全穀物中的膳食纖維和植物化學物，如燕麥中的β-葡聚醣和燕麥蒽胺，有助於降低促炎細胞因數和炎症生物標記物水準，抑制腸道病原體生長，通過對腸道和腸道相關的淋巴組織產生有利影響，增加機體免疫功能，促進機體整體健康水準。

綜上，全穀物中含有多種有益健康的營養素（包括膳食纖維、多種植物營養素、維他命、礦物質以及不飽和脂肪酸），每天安排至少一頓含有全穀物的主食，將對健康產生積極且深遠的影響。

4 Okarter, N. and R.H. Liu, Health Benefits of Whole Grain Phytochemicals[J]. Critical Reviews in Food Science & Nutrition, 2010 (No.3): 193-208

5 Okarter, N, et al, Phytochemical content and antioxidant activity of six diverse varieties of whole wheat[J]. Food Chemistry, 2010.119(1): 249-257.

6 中國營養學會營養健康研究院. 健康穀物白皮書[J].2021-1.

常見的全穀物及其製品

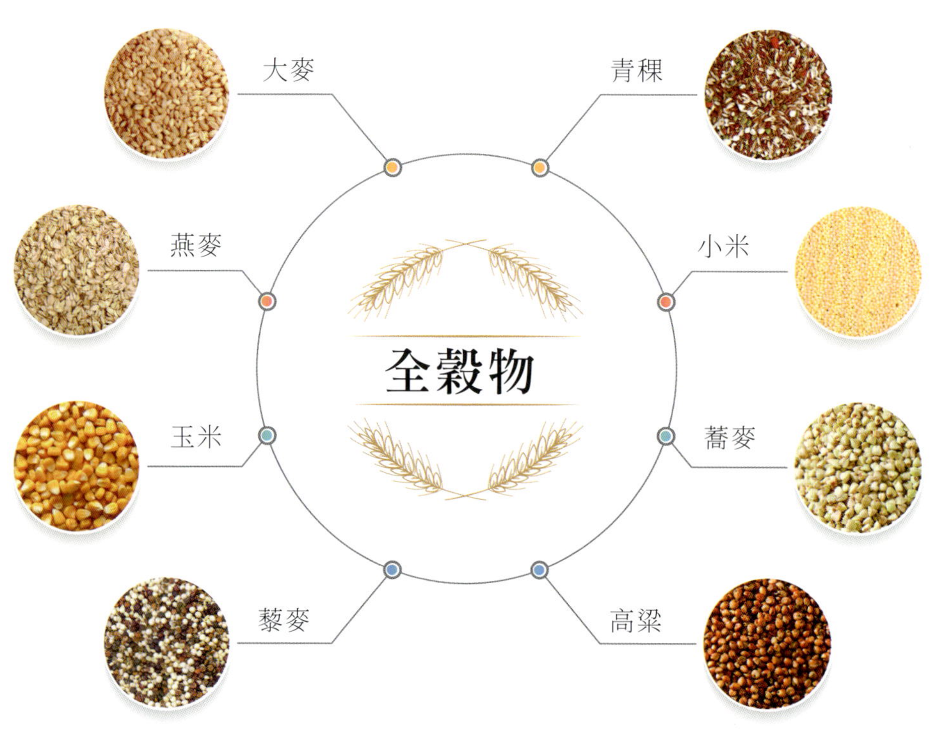

常見的全穀物製品

全麥麵包

雜糧饅頭

雜糧飯

糙米粥

即食燕麥片

混合穀物粉

如何選購全穀物製品

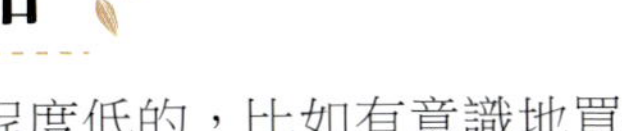

在購買主食時，我們應該購買加工程度低的，比如有意識地買一些糙米、全麥產品等。這時閱讀食品標籤就非常重要。

1. 查閱標籤中是否含有「全麥」、「全穀物」、「全麥粉」等字樣。選擇標籤上注明全穀物百分比含量的製品。

2. 確認全穀物原料在食品配料表中的排位或添加量。優選全穀物在配料表中排第一或第二項的。

看營養成分表，同款產品儘量選膳食纖維含量高的。國家規定，如果每 100g 產品中至少含 6g 膳食纖維，就可以稱之為「高或富含膳食纖維」。如果主食中膳食纖維含量達到這個標準，全穀物的添加量通常也不會低。

燕麥是一種全穀物

燕麥是禾本科

燕麥屬一年生草本植物，一般分為帶稃型（皮燕麥）和裸粒型（裸燕麥）兩種。我國是皮燕麥的故鄉，遠古時代就是我國人民的主糧之一。燕麥是營養價值很高的糧飼兼用作物，世界範圍內其種植面積僅次於小麥、水稻、玉米，居於糧食作物的第四位。燕麥在加工過程中，幾乎不會丟失其麩皮和胚芽，屬於一種全穀物。作為全穀物的一份子，燕麥具備該類別食物的共性，同時自身也具備。

燕麥 *Old oats*

100%純天然燕麥片，煮5分鐘即可。

鋼切燕麥 *Steel-cut oats*

將燕麥簡單切塊，未經壓扁卷製處理，需要長時間煮制。雖耗時，但口感飽滿、味道濃郁。

即食燕麥 *Instant oatmeal*

也叫速食燕麥。經過預先煮熟和用熱水沖製、或快煮1~2分鐘即可食用。加工程度高，讓燕麥的質地和結構大大流失，所以烹製成品並不會保留太多燕麥的質地和口感，適合忙碌的學生、上班族等生活節奏快的人群。

燕麥製品 *Oat products*

以燕麥為原材料之一的食品，包括風味燕麥片、燕麥飲料、燕麥麵包等。

快熟燕麥 *Quick-cook oatmeal*

經過先蒸後壓製而成，煮製所需時長比傳統燕麥長，比鋼切燕麥短。烹製成品一定程度保留了燕麥的質地和口感。

燕麥的營養特點

燕麥是一種營養價值豐富的食物，是膳食纖維、礦物質、維他命B雜以及植物化學物（包括：酚類、β-葡聚醣、燕麥生物鹼、木質素以及植物甾烷醇）的良好來源。燕麥β-葡聚醣是燕麥的特徵性營養成分，其含量佔3~6%，是有益心血管、血糖和腸道健康的活性成分，對腸道菌群也能產生有利影響。燕麥主要被人們以燕麥片、穀物早餐以及燕麥麵粉的形式食用。

燕麥營養成分（每100g可食部）	蛋白質	維他命B_1	維他命B_2	煙酸	鐵	鋅	膳食纖維
	16.9g	0.8mg	0.1mg	1.0mg	4.7mg	4.0mg	10.6g

*數據來源：美國農業部資料庫

燕麥的健康益處

燕麥作為一種近年來被廣泛關注的農作物，已經越來越引起營養學界的興趣，關於燕麥科學、技術和健康效應的研究層出不窮。幾十年間，大量的研究表明燕麥具有健康益處，促使一些國家政府批准與燕麥相關的健康聲稱，包括加拿大、美國、澳大利亞、馬來西亞等 —— 燕麥是第一種獲得政府認證的健康食物，主要得益於其特有的β-葡聚醣成分。

有益心臟健康：燕麥特有的β-葡聚醣可有效降低血液中的總膽固醇和低密度脂蛋白膽固醇（LDL）水準，從而降低心血管疾病的發生風險。研究證實，每日攝入至少3g燕麥β-葡聚醣（約75g燕麥）能使血膽固醇水準降低。[7]

有益血糖平衡控制：燕麥β-葡聚醣可增加食糜黏度，使得胃排空速度減慢，降低機體血糖應答反應。[8] β-葡聚醣在腸道內被菌群發酵及代謝生成的短鏈脂肪酸，會進入到人體血液中，被轉運到其他器官，間接影響血糖。

有益腸道健康：燕麥能提供可溶性膳食纖維、抗性澱粉以及低聚醣，可促進排便、預防便秘。燕麥中的β-葡聚醣不能被人體消化吸收，攝入後能有效發揮益生元的作用。[9]

總而言之，燕麥的促健康效應得到越來越多的證實，新的健康益處被不斷地發現，其背後的潛在機制被一步步地揭開。對於普通消費者來說，我們不需要深入瞭解這麼專業的資訊，只要從認知上瞭解燕麥有益健康，行動上選對了產品 —— 無過多其他成分添加的純燕麥產品，再加上規律的鍛煉，就能讓全家人享受到更大的健康福利。

7 向雪松，孫建琴，葉夢瑤等。燕麥對血膽固醇邊緣性升高人群血脂水準的影響：一項隨機對照研究[J]。營養學報，2019,41(03):242-247。

8 孟彥彤，張東傑，薛勇等。燕麥β-葡聚醣的功效研究進展[J/OL]。中國糧油學報:1-16[2024-01-04]。

9 李程，邢青斌，孫桂菊，向雪松。燕麥／大麥β-葡聚醣的益生元效應研究證據[J]。營養學報，2022, 44 (04): 404-409。

燕麥 β-葡聚醣益生元效應

β-葡聚醣是一種多醣類物質，廣泛存在於植物和微生物中。不同來源的β-葡聚醣根據結構的差異，在水溶性和功能側重點上有所不同。燕麥中的β-葡聚醣為植物來源，其結構多為線性β-（1,3；1,4）[10]-D-葡聚醣。中國於2014年已批准燕麥β-葡聚醣為新食品原料，並且燕麥β-葡聚醣也通過了美國FDA的GRAS認證。

β-葡聚醣無法被人體胃及小腸中的消化酶分解，因此能夠進入結腸，被特定腸道菌群利用並代謝。近年來，燕麥β-葡聚醣的益生元效應已被多個高質量的人體科學研究證實。

1.燕麥β-葡聚醣能夠“被宿主微生物選擇性利用”[11]：

能顯著促進有益菌的增殖，如雙歧桿菌、乳桿菌等；並能促進腸道菌群相關代謝產物如短鏈脂肪酸生成，進而促進腸道健康。

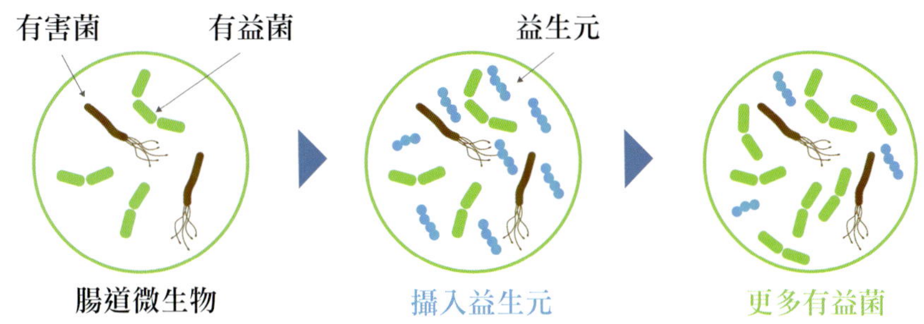

2.燕麥β-葡聚醣有助於降低血清膽固醇濃度：

β-葡聚醣可通過調節腸道特定菌群如羅氏菌屬和布勞特氏菌屬等菌群數量進而影響機體脂質代謝。[12]研究結果顯示2.9～5.0g/d β-葡聚醣能夠顯著降低TC和LDL-C含量。美國食物藥品監督管理局、加拿大和歐洲等國家和地區均授權了針對β-葡聚醣攝入與降低心血管疾病風險的健康宣稱。

3.燕麥β-葡聚醣有助於抑制餐後血糖上升[13]：

β-葡聚醣經過腸道菌群發酵後產生的SCFAs可通過調節胰島素依賴的促胰島素多肽（GIP）和胰高血糖素樣肽-1（GLP-1）等內分泌激素水準來抑制食慾、提高胰島素敏感性和胰島素水準，發揮血糖調節作用。

10 Gibson GR, Hutkins R, Sanders ME, et al. Expert consensus document： The International Scientific Association for Probiotics and Prebiotics (ISAPP) consensus statement on the definition and scope of prebiotics [J]. Nat Rev Gastroenterol Hepatol, 2017, 14： 491－502.

11 Korczak R, Kocher M, Swanson KS. Effects of oats on gastrointestinal health as assessed by in vitro, animal, and human studies [J]. Nutr Rev, 2019, 78： 343－363.

12 Xu D, Feng M, Chu Y, et al. The prebiotic effects of oats on blood lipids, gut microbiota and short-chain fatty acids in mildly hypercholesterolemic subjects compared to rice: a randomized, controlled trial [J]. Front Immunol, 2021， 12： 787－797.

13 Steinert RE, Raederstorff D, Wolever TM. Effect of consuming oat bran mixed in water before a meal on glycemic responses in healthy humans-a pilot study [J]. Nutrients, 2016, 8： 524.

將燕麥融入一日三餐

健康食品也能做得美滋美味！人們常將健康食品與索然無味畫上等號，大家也許會秉持「健康食物難以下咽」的一貫邏輯，但事實遠非如此！有多種方法為菜餚添香增色。例如，燕麥是一種多用途的食材，而且燕麥的用途廣泛，能輕鬆且巧妙地融入到一日三餐以及加餐中，讓您的菜品味道更豐富、口感更多樣，讓就餐體驗變得驚喜多多、美味多多！

將燕麥納入膳食的烹飪方式多種多樣，簡單舉幾個例子：

燕麥烘焙食物
如燕麥蛋糕、燕麥麵包以及燕麥曲奇餅乾等 。

隔夜燕麥水果/堅果撈
將燕麥放入碗中，倒入牛奶、優酪乳和水果，放入冰箱浸泡一整夜。

燕麥新式吃法
燕麥厚蛋燒、燕麥可麗餅、豆乳桂花燕麥撻等，利用燕麥做出新式菜譜。

燕麥優酪乳堅果隨心搭
將燕麥放到紫薯、紅薯、南瓜等泥中，澆一杯優酪乳，加一些乾果及堅果碎兒，攪拌均勻後食用。

燕麥傳統主食
將燕麥加入到麵粉中，做成燕麥饅頭或者麵條或雜糧飯食用。

燕麥水果沙拉或布丁

■此處只列舉了少數幾種美食做法，將燕麥融入每日膳食當中製作的美味還有很多，篇幅限制無法在此一一列出，趕緊翻開本書後面的章節一探究竟吧！

QUAKER
桂格
EST? 1877

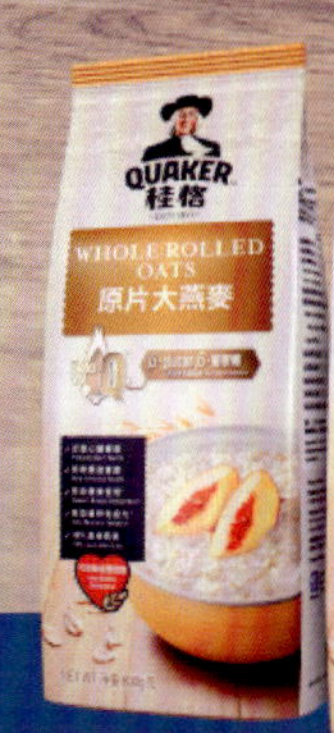
QUAKER
桂格
WHOLE ROLLED OATS
原片大燕麥

QUAKER
桂格
INSTANT OATMEAL
即食燕麥片

QUAKER
桂格
QUICK COOK OATMEAL
快熟燕麥片

早餐系列
Breakfast

大蝦燕麥粥
食材份量 / Ingredient
早餐系列
Breakfast
食油 3ml/毫升
Cooking Oil
酒 5ml/毫升
Cooking Wine
雞湯 210ml/毫升
Chicken Soup
生抽 3ml/毫升
Soy Sauce
原隻大蝦
Prawns 2pcs/隻
麻油 1ml/毫升
Sesame Oil
香菜（芫荽）
Coriander 3g/克
水 50ml/毫升
Water
薑 2g/克
Ginger
桂格快熟燕麥片 30g/克
Quick Cook Oatmeal
胡椒粉
Pepper 1g/克
QUAKER
桂格
QUICK COOK
OATMEAL
快熟燕麥片

Prawns Oat Congee
大蝦燕麥粥

製作步驟 / Step

1 鍋內倒油，油熱後放入大蝦炒熟備用。
Spray oil over pan, stir fried prawns once the oil is heated, then set aside.

2 炒熟大蝦後加入燕麥片、料酒、雞湯、薑絲、胡椒粉，加水煮開，煮至收汁時便可。
Add oatmeal, cooking wine, chicken soup, shredded ginger, pepper and water with cooked prawns, boiled till sauce thickens and absorbed.

3 放香菜裝飾，淋上麻油和醬油。
Garnish with coriander, drizzle with sesame oil and soy sauce before serving.

優點及營養成份

大蝦燕麥粥是一款營養豐富、美味可口的粥品，它以大蝦和燕麥為主要食材。大蝦富含優質蛋白質、維他命和礦物質等營養素。燕麥是一種低GI值（升糖指數）、高纖維的穀物，有益於促進腸道健康。本款燕麥粥口感豐富，質地軟糯，是一道營養豐富的快手餐。

香菇雞肉燕麥粥

早餐系列

Breakfast

食材份量 / Ingredient

玉米粒 30g/克
Corn Kernels

食油 3ml/毫升
Cooking Oil

香菇 2pcs/隻
Mushroom

鹽 0.5g/克
Salt

水 320ml/毫升
Water

雞胸肉 40g/克
Chicken Breast

桂格即食燕麥片 40g/克
Instant Oatmeal

Shiitake and Chicken Oat Congee
香菇雞肉燕麥粥

製作步驟 / Step

1 香菇切片。
Slice mushroom.

2 雞胸肉切粒。
Dice chicken breast into pieces.

3 鍋中放入食油燒熱，加入香菇和雞胸肉，中火不斷翻炒至變色，再加入水320毫升。
Spray oil over pan, add shiitake mushroom and chicken breast, maintain medium heat and stir fried until the color changes, then add 320ml water.

4 煮沸後加入玉米粒、燕麥片，收汁至稠。
Add corn kernels and oatmeal, boil till sauce thickens and absorbed.

5 加鹽調味即可。
Season with salt before serving.

優點及營養成份

香菇雞肉燕麥粥中蛋白質及膳食纖維含量均十分豐富。本食式中採用的桂格即食燕麥片，口感軟糯，不僅適合成年人，老人小孩也可以食用。

芋香燕麥粥

食材份量 / **Ingredient**

早餐系列
Breakfast

玉米粒 30g/克
Corn Kernels

食油 2ml/毫升
Cooking Oil

葱 20g/克
Green Onion

松茸 20g/克
Matsutake

鹽 0.5g/克
Salt

芋頭
Taro 70g/克

桂格快熟燕麥片 30g/克
Quick Cook Oatmeal

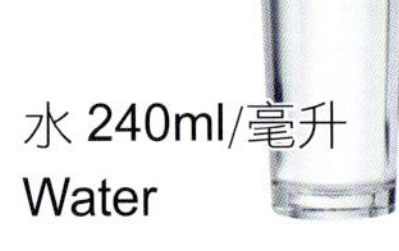

水 240ml/毫升
Water

Taro Oat Congee
芋香燕麥粥

製作步驟 / Step

1 將燕麥片、玉米粒切好的芋頭放入電飯煲，加入水，按煮粥鍵1小時。

Put oatmeal, corn kernels, and chopped taro into rice cooker. Add water and press "cook porridge/gruel/congee" for 1 hour.

2 鍋內放油，油熱後煎松茸，煎至兩面金黃。

Spray oil over pan, fried matsutake until both sides are golden brown.

3 將煮好的粥放入鹽調味，放上煎好的松茸，最後撒上蔥花。

Season the porridge with salt, place fried matsutake on top and sprinkle with chopped green onion before serving.

優點及營養成份

芋頭口感綿甜香糯，味道濃香，顏色呈淺紫色，和燕麥一起製作成粥品更加香濃且易於消化。松茸中含有的松茸多醣和 β- 葡聚醣可以幫助維持免疫力。與傳統粥品相比較，芋香燕麥粥營養價值更高，製作簡單快捷，很適合上班族。

番茄雞蛋燕麥粥

早餐系列
Breakfast

食材份量 / Ingredient

雞蛋 1pc/隻
Egg

番茄 1pc/隻
Tomato

生菜 30g/克
Lettuce

食油 6ml/毫升
Cooking Oil

生抽 3ml/毫升
Soy Sauce

桂格即食燕麥片 50g/克
Instant Oatmeal

水 400ml/毫升
Water

Tomato and Egg Oat Congee
番茄雞蛋燕麥粥

製作步驟 / Step

1 雞蛋打成蛋液，番茄切小塊，生菜切碎。
Beat eggs until thoroughly mixed, dice tomato and cut lettuce into stripes.

2 鍋中加食用油，把雞蛋炒熟，盛入碗中備用。
Spray oil over pan, scramble the egg and set aside.

3 番茄炒至軟淋，加水燒開後放入燕麥片和炒熟的雞蛋，煮至收汁至稠。
Fry tomato until soften, add water, oatmeal, and the scramble egg and boil till sauce thickened and absorbed.

4 煮熟後放入生抽調味。
Season with soy sauce.

5 最後放生菜裝飾。
Garnish with lettuce before serving.

優點及營養成份

番茄燕麥粥結合了番茄、雞蛋和燕麥三種食材，用燕麥片來代替傳統的大米，不僅口感獨特，而且營養價值較高。同時，燕麥中含蛋白質和膳食纖維，此外，粥品製作起來也比較簡單，是一道健康佳餚選擇。

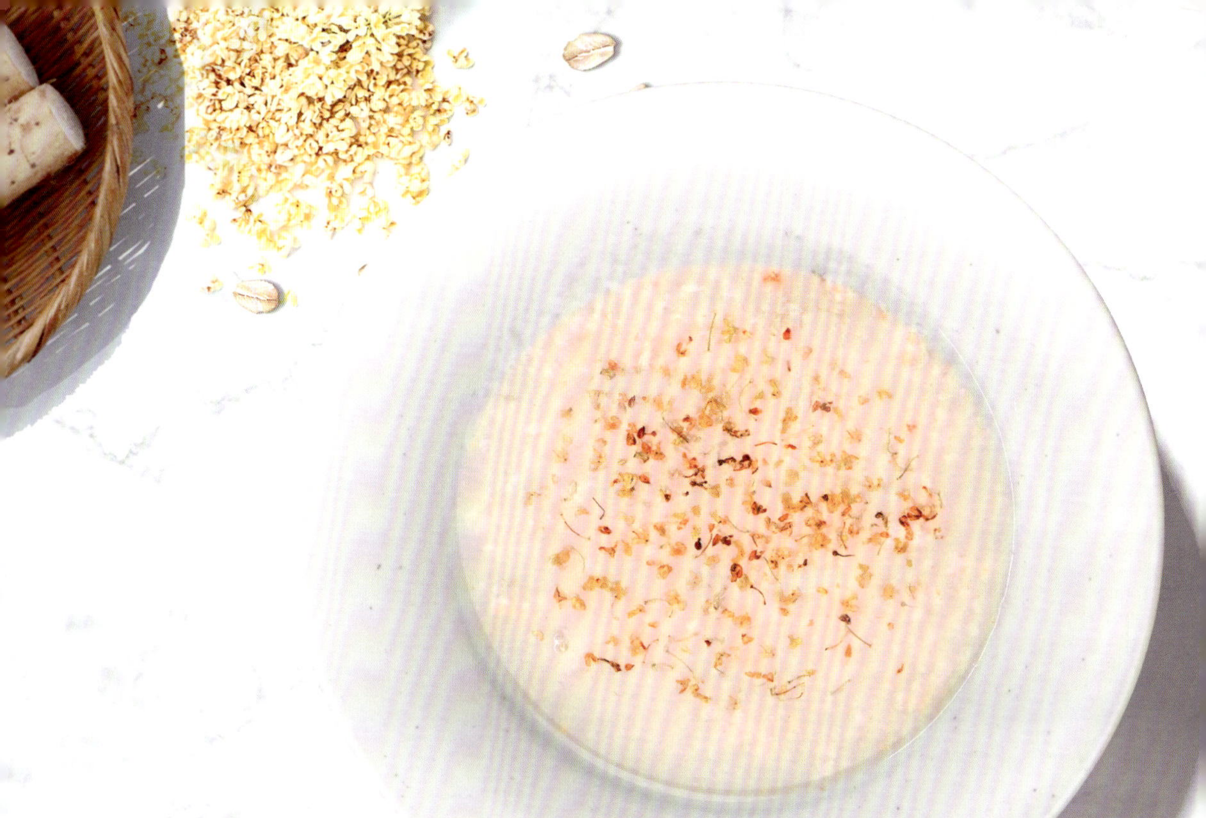

淮山桂花燕麥粥

早餐系列 *Breakfast*

食材份量 / Ingredient

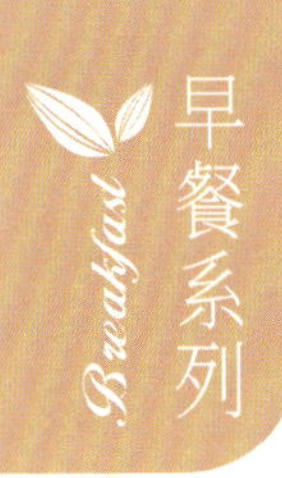

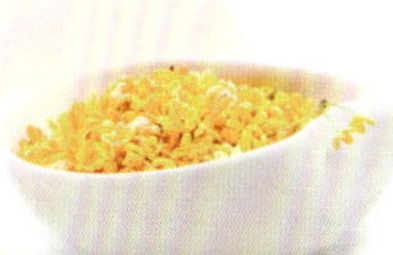

桂花 10g/克
Osmanthus

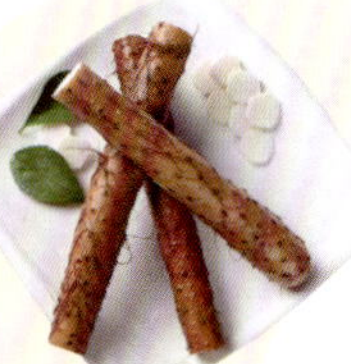

淮山 50g/克
Chinese Yam

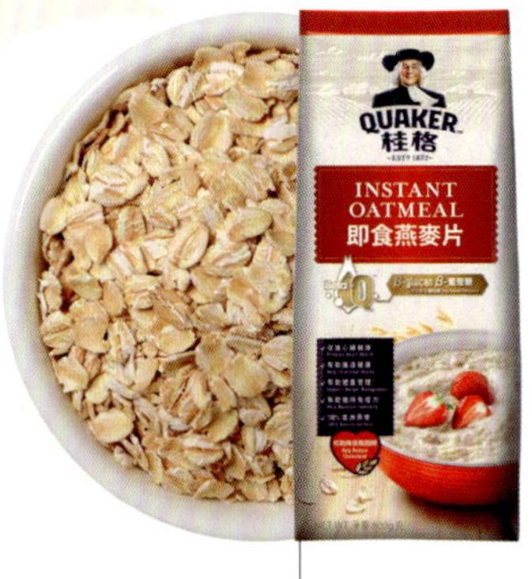

桂格即食燕麥片 40g/克
Instant Oatmeal

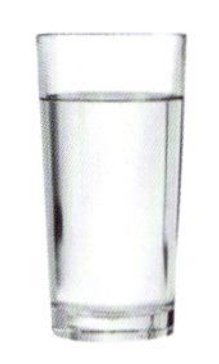

水 240ml/毫升
Water

Chinese Yam and Osmanthus Oat Congee
淮山桂花燕麥粥

製作步驟 / Step

1 淮山去皮切成塊。
Peel Chinese yam and cut it into pieces.

2 切好的淮山放入鍋裏，加240毫升清水，大火煮開轉小火，煮至淮山熟透。
Place the chopped Chinese yam into pot, add 240ml water and boil it over high heat, once it boiled, reduce to low heat and cook the Chinese yam thoroughly.

3 放入燕麥片和桂花，混合攪拌。
Mix the oatmeal and osmanthus well.

4 蓋上鍋蓋燜5分鐘即可食用。
Cover with the lid and wait for 5 minutes before serving.

優點及營養成份

淮山中含有豐富的澱粉酶、多酚等物質，可以促進胃腸道消化吸收功能。搭配軟糯、營養豐富的桂格即食燕麥片，更易消化吸收。此外，桂花自帶的香氣使樸實的白粥多了一絲清甜的味道。

時蔬蝦仁燕麥粥

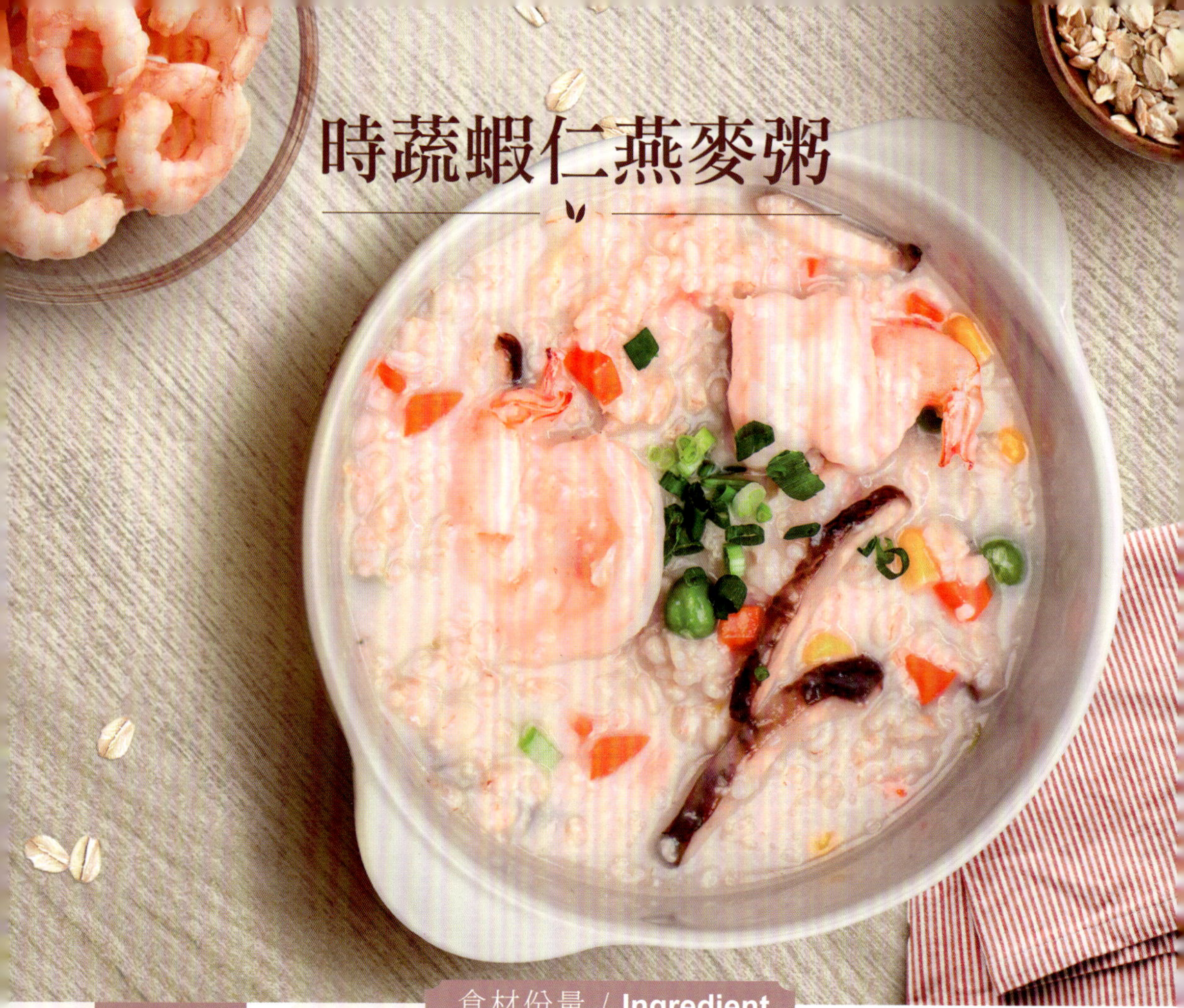

食材份量 / Ingredient

早餐系列 *Breakfast*

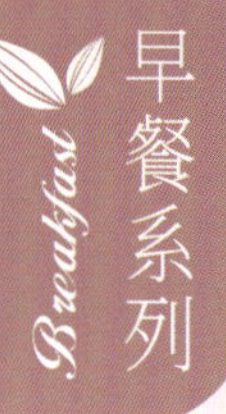

蝦仁 4pcs/隻
Prawns

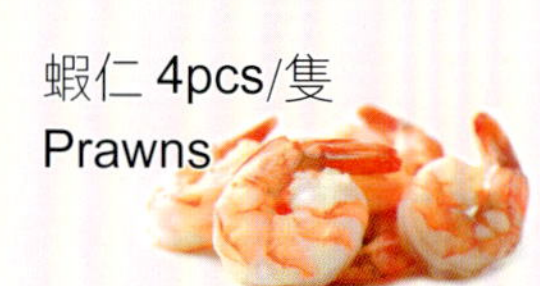

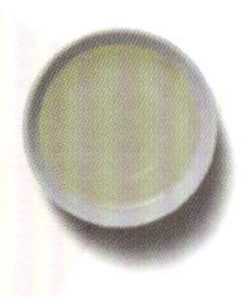

食油 3ml/毫升
Cooking Oil

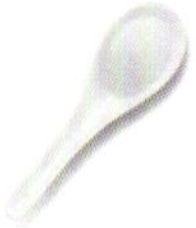

鹽 0.5g/克
Salt

玉米粒 30g/克
Corn Kernels

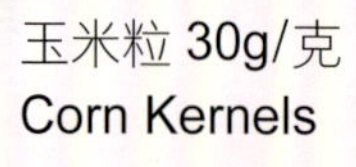

青豆 30g/克
Green Bean

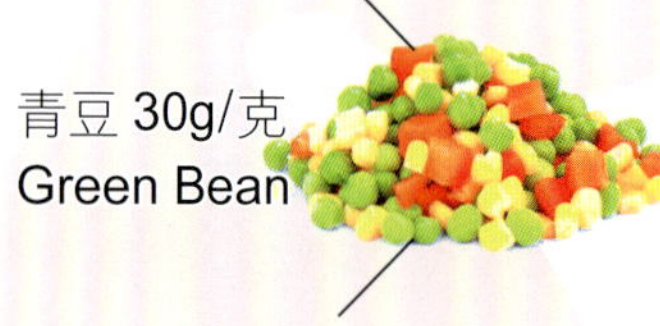

胡蘿蔔 30g/克
Carrot

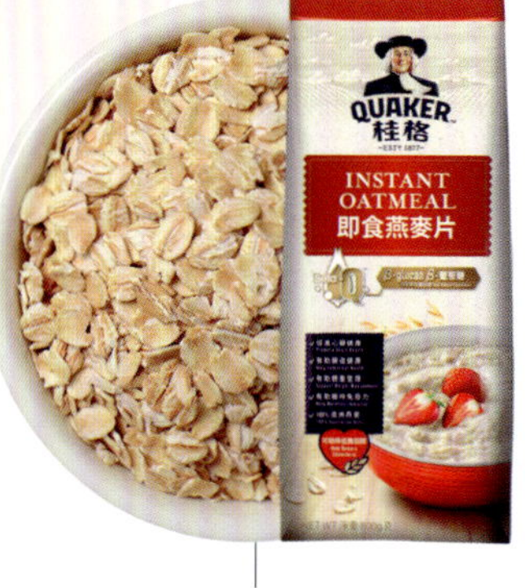

桂格即食燕麥片 50g/克
Instant Oatmeal

水 50ml/毫升
Water

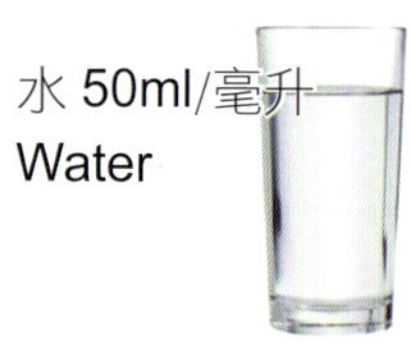

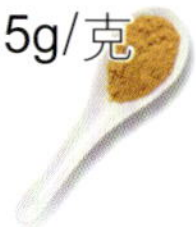

胡椒粉 0.5g/克
Pepper

Vegetable and prawns Oat Congee
時蔬蝦仁燕麥粥

製作步驟 / Step

1 鍋中放入食用油，油熱後放入蝦仁炒至變色。
Spray oil over pan, stir fried the prawns until the color changes.

2 放入青豆、玉米粒、胡蘿蔔，與蝦仁一起翻炒出香味。
Add the green bean, corn kernels and carrot, and stir-fry together with the prawns.

3 鍋中加入清水，煮5分鐘。
Add water to the pan and cook for 5 minutes.

4 放入燕麥片（確定水剛好沒過燕麥片，如水不夠可再次加入少許清水）後再煮5分鐘。
Add the oatmeal (make sure it is covered by water, you may add more water to it if needed), then cook for another 5 minutes.

5 加入胡椒粉及鹽調味即可。
Season with pepper and salt before serving.

優點及營養成份

燕麥富含膳食纖維，代替傳統熬粥用的精製白米，加上青豆、玉米粒、胡蘿蔔等食材共同熬製成粥，可幫助調整升糖指數（GI）而且含有更多的營養素和膳食纖維，粥中蝦仁屬於優質蛋白食物，可作為早餐的動物優質蛋白質來源，具有低脂肪、高蛋白質的特點。

酒釀雞蛋燕麥粥

早餐系列
Breakfast

食材份量 / Ingredient

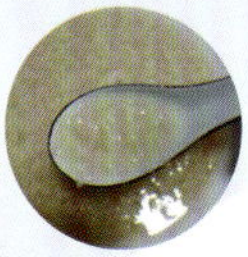

酒釀(糟鹵) 20g/克
Fermented Rice Wine

雞蛋 1pc/隻
Egg

黑芝麻 0.5g/克
Black Sesame

牛奶 150ml/毫升
Milk

桂花 10g/克
Osmanthus

桂格即食燕麥片 25g/克
Instant Oatmeal

Oat Congee With Poached Egg in Sweet Fermented Rice wine
酒釀雞蛋燕麥粥

製作步驟 / Step

1 打一隻雞蛋攪拌備用，牛奶裏加入酒釀(糟鹵) 煮開。
Beat an egg and set aside. Bring the milk to a boil, then add the fermented rice wine and stir.

2 煮開後，加入燕麥片，煮3-5分鐘。
Add the oatmeal after boiling and cook for 3-5 minutes.

3 鍋內加入雞蛋液，攪拌均勻。
Stir in beaten egg.

4 完成後撒上桂花和黑芝麻點綴。
Garnish with osmanthus fragrans and black sesame before serving.

優點及營養成份

酒釀雞蛋燕麥粥中食材十分豐富，其中的牛奶鈣含量豐富，同時燕麥本身含有豐富的膳食纖維，有助於增強飽腹感、有利於腸道健康。再搭配上桂花自帶的香氣和酒釀（糟鹵）的醇香，是一道美味粥品。

南瓜燕麥蒸糕

食材份量 / **Ingredient**

早餐系列 *Breakfast*

低筋麵粉 60g/克
Low Gluten Flour

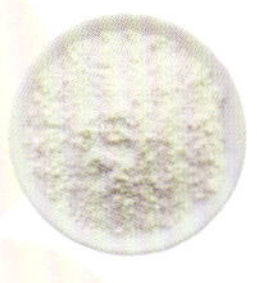

南瓜泥 40g/克
Pumpkin Puree

葡萄乾 4g/克
Raisin

泡打粉 1.5g/克
Baking Powder

蔓越莓乾 4g/克
Dried Cranberry

乾酵母 1.5g/克
Dry Yeast

牛奶 30ml/毫升
Milk

砂糖 25g/克
Sugar

桂格即食燕麥片 40g/克
Instant Oatmeal

Steamed Pumpkin Oat Cake
南瓜燕麥蒸糕

製作步驟 / Step

1 留出少量蔓越莓和葡萄乾，將剩餘原料加在一起，拌勻即可。
Reserve some dried cranberries and raisins for later use. Mix the remaining ingredients together with the dough.

2 放入模具，醒發4小時（20°C-25°C之間），撒上少許蔓越莓和葡萄乾，蒸烤箱100°C蒸15-20分鐘。
Place the dough in a mold and let it rise for 4 hours at 20°C-25°C. Garnish with the reserved dried cranberries and raisins, then steam it at 100°C for 15-20 minutes.

優點及營養成份

蒸糕是一道傳統的點心，以麵粉為主要原料，口感軟糯、香甜可口，將燕麥和南瓜混合到蒸糕中，既增加了蒸糕的口感，也提高了營養價值。此外，蔓越莓和葡萄乾的點綴也為蒸糕增添果香和甜味。這款南瓜燕麥蒸糕適合想控制油、鹽、糖攝取的人食用，是一道健康又美味的點心。

燕麥片糯米餅

食材份量 / Ingredient

糯米粉 50g/克
Glutinous Rice Flour

葵花籽油 3ml/毫升
Sunflower Oil

桂格即燕食麥片 25g/克
Instant Oatmeal

水 35ml/毫升
Water

Sticky Rice and Oat Cake
燕麥片糯米餅

製作步驟 / Step

1 將燕麥片混入糯米粉中，加35毫升清水揉成光滑的燕麥片糯米團。
Mix the oatmeal with glutinous rice flour. Add 35ml water and knead the dough until the surface is smooth.

2 將燕麥片糯米團分成大小均勻的小麵團，搓圓按扁，並整理成圓形。
Divide the dough into equal portions and shape each into a round flat shape.

3 平底鍋中倒入葵花籽油，放入做好的燕麥片糯米餅的煎至表面焦香即可。
Spray sunflower oil over pan and fry it until both sides are golden brown.

優點及營養成份

糯米含有豐富的碳水化合物及維他命 B 雜，作為早餐食用，可以快速補充能量。燕麥富含膳食纖維，鐵、鎂等礦物質。燕麥與糯米搭配食用，既可以獲得多種營養素，又能豐富食物口感。

燕麥鱈魚餅

食材份量 / **Ingredient**

早餐系列 *Breakfast*

鱈魚 50g/克
Codfish

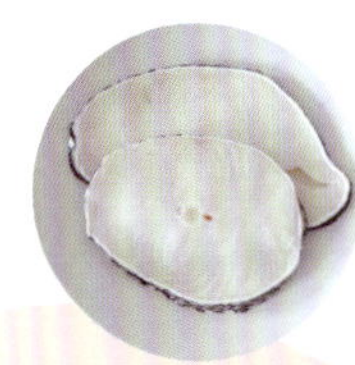

澱粉 5g/克
Starch

南瓜 100g/克
Pumpkin

橄欖油 5ml/毫升
Olive Oil

桂格即食燕麥片 30g/克
Instant Oatmeal

水 10ml/毫升
Water

Codfish Oat Cake 燕麥鱈魚餅

製作步驟 / Step

1 南瓜放於鍋中，蒸10分鐘至熟透，搗成泥備用。
Steam pumpkin for 10 minutes, then mash it and set aside.

2 鱈魚、燕麥片、10毫升水及5克澱粉用攪拌機中攪拌成泥。
Add the codfish, oatmeal, 10ml water, blend it into a puree and 5g starch into mixer.

3 將鱈魚燕麥泥與南瓜泥一起放於碗中，攪拌混勻。
Mix codfish oatmeal puree with pumpkin mash.

4 平底鍋中刷上橄欖油，燒熱後，將步驟3中的食材平攤在鍋中，小火煎至定型，翻面後繼續煎。
Spray olive oil over pan. Fry the mixture from step 3 over low heat. Flip and cook the other side until both sides are golden brown.

5 繼續小火煎至雙面焦香、熟透即可取出食用。
Continue frying over low heat until both sides are golden brown.

優點及營養成份

燕麥鱈魚餅的膳食纖維及蛋白質含量較高，是一道高營養價值的菜品。鱈魚富含DHA，尤其適合孕產婦、哺乳期婦女、兒童食用。南瓜本身具有的甜味和鱈魚的鹹味，不需要額外添加鹽、糖等調味品，是一道相對健康美味的快手菜。

太陽燕麥餅

食材份量 / Ingredient

早餐系列
Breakfast

紅薯 80g/克
Sweet Potato

雞蛋清 1pc/隻
Egg White

櫻桃番茄 10g/克
Cherry Tomato

桂格5紅混合即食燕麥片 40g/克
5 Red Multi-Grain

Sweet Potato Oat Pancakes
太陽燕麥餅

製作步驟 / Step

1 紅薯洗淨烤熟或煮熟，去皮放入碗裏搗碎，加入燕麥片拌勻。

Boil or bake the sweet potato after cleaning. Remove the skin and mash it into puree. Add oatmeal and mix it well.

2 按壓成燕麥薯餅。

Press and shape the mixture.

3 蛋清打發，用勺舀在餅底上，烤箱190°C，烤15分鐘。放櫻桃番茄或堅果點綴。

Beat the egg white and place it under the cake mixture, bake at 190°C for 15 minutes. Garnish with cherry tomatoes or nuts before serving.

優點及營養成份

紅薯屬於薯類，含有豐富的維他命，其中 β- 胡蘿蔔素和維他命 E 尤為豐富。燕麥含膳食纖維。太陽燕麥餅中的食材豐富，營養全面，口感鬆軟，小朋友和老人均可食用。

芝士蟹柳燕麥可麗餅

早餐系列
Breakfast

食材份量 / Ingredient

檸檬汁（適量）
Lemon Juice a pinch

雞蛋 1pc/隻
Egg

食油 2ml/毫升
Cooking Oil

蟹柳 1pc/條
Crab Stick

胡椒粉 0.5g/克
Pepper

牛油果（適量）
Avocado a pinch

鹽 0.3g/克
Salt

低脂芝士片 1/2pc/片
Low Fat Cheese

水 80ml/毫升
Water

生菜（適量）
Lettuce a pinch

桂格即食燕麥片 55g/克
Instant Oatmeal

Cheese Crabstick Oat Crepe
芝士蟹柳燕麥可麗餅

製作步驟 / Step

1 碗中加入燕麥片、雞蛋和80毫升清水攪拌均勻。
Mix the oatmeal, egg and 80ml water in a bowl.

2 平底鍋中倒入油，燒熱後倒入燕麥糊，小火煎燕麥餅熟後備用。
Spray oil over pan, pour mixture into the heated pan, and cook over low heat and then set aside.

3 把牛油果放入碗中搗成泥，滴入幾滴檸檬汁攪拌均勻。
Puree the avocado and mix with a few drops of lemon juice.

4 燕麥餅上放上生菜、低脂芝士片，塗抹牛油果泥。
Place lettuce, low fat cheese and avocado puree on the oatmeal crepe.

5 將蟹柳撕成絲平鋪在上方，撒胡椒粉、鹽調味即可。
Tear the crab stick into stripes and place it on top of the crepe. Season it with pepper and salt before serving.

優點及營養成份

與傳統可麗餅相比，使用燕麥代替麵粉，並避免額外添加糖和黃油製作的燕麥可麗餅，具有較低 GI 以及含膳食纖維的特點。其餡料由蟹柳和低脂芝士片這兩種低脂高蛋白食材搭配而成。此外，由於避免了使用蜂蜜、巧克力醬等高添加糖食材，使得整個菜餚營養素相對更健康全面。

韭菜燕麥雞蛋餅

早餐系列
Breakfast

食材份量 / Ingredient

雞蛋 1pc/隻
Egg

食油 2ml/毫升
Cooking Oil

麵粉 10g/克
Flour

生抽 3ml/毫升
Soy Sauce

韭菜 （適量）
Chinese Chives a pinch

桂格即食燕麥片 25g/克
Instant Oatmeal

胡蘿蔔 10g/克
Carrot

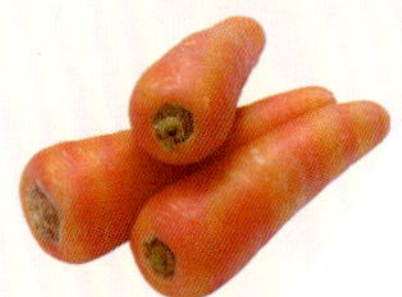

水 80ml/毫升
Water

Chive and Egg Oat Pancake
韭菜燕麥雞蛋餅

製作步驟 / Step

1 水加熱後，加入燕麥片，漲發5分鐘後加入麵粉拌勻，用盤子盛起。
Pour hot water over the oatmeal and let it soak for 5 minutes, then mix it well with flour.

2 在煮熟的燕麥片裏面加入雞蛋、韭菜、胡蘿蔔、醬油攪勻。
Add egg, Chinese chives, carrot and soy sauce to the cooked oatmeal mixture.

3 煎鍋裏加食用油，油燒熱後把拌好的燕麥片倒入鍋裏煎至兩面熟透即可。
Spray oil over pan and fry it until both sides are golden brown.

優點及營養成份

這款韭菜燕麥雞蛋餅在麵粉和雞蛋的基礎上，添加了富含 β- 胡蘿蔔素的胡蘿蔔和富含膳食纖維即食燕麥片，以維持飽腹感。韭菜的香氣使得餅的味道更加豐富，烹飪完成後香氣四溢。

五黑燕麥煎餅卷

早餐系列
Breakfast

食材份量 / **Ingredient**

牛肉 30g/克
Beef

全麥捲餅皮 1pc/片
Whole Wheat Wrap

彩椒 （適量）
Bell Pepper a pinch

熟鹹蛋黃 1.5pcs/個
Cooked Salted Egg Yolk

鹽 0.5g/克
Salt

牛油果 20g/克
Avocado

黑胡椒 0.2g/克
Black Pepper

生菜（適量）
Lettuce a pinch

桂格5黑混合即食燕麥片 40g/克
5 Black Multi-Grain

熱水 120ml/毫升
Hot Water

Five Black Multi-Grain Pancake Roll
五黑燕麥煎餅卷

製作步驟 / Step

1 將牛肉撒上黑胡椒和鹽烤熟，牛油果用勺子壓成泥，彩椒切條，生菜切絲。
Season the beef fillet with pepper and salt. Mash the avocado to puree, slice bell pepper and chop lettuce for later use.

2 將燕麥片用熱水泡發5分鐘，待用。
Soak the oatmeal in hot water for 5 minutes.

3 起鍋（平底不粘鍋），放入全麥卷餅皮，小火加熱。
Heat the whole wheat wrap in a dry non-stick pan over low heat until warm.

4 全麥卷餅皮上放泡發的燕麥片、熟鹹蛋黃、烤牛肉、生菜、彩椒，塗抹牛油果泥，捲起。
Place the soaked oatmeal, cooked salted egg yolk, grilled beef fillet, lettuce, bell pepper and avocado puree on the whole wheat wrap before wrapping.

優點及營養成份

一份食材豐富的五黑燕麥煎餅卷，滿滿穀物香氣加上牛肉與果蔬的豐富口感包裹著味蕾，幾分鐘就能做出的養生健康花式早餐，非常適合減脂的人食用。

燕麥全麥饅頭

早餐系列
Breakfast

食材份量 / **Ingredient**

麵粉 10g/克
Flour

泡打粉 0.6g/克
Baking Powder

溫水 8ml/毫升
Warm Water

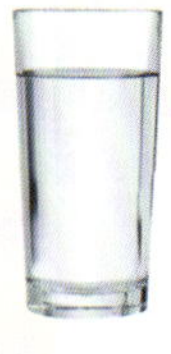
葵花籽油 2ml/毫升
Sunflower Oil

桂格即食燕麥片 4g/克
Instant Oatmeal

酵母 0.8g/克
Yeast

砂糖 25g/克
Sugar

水 20ml/毫升
Water

Steamed Whole-Wheat Oat Bun
燕麥全麥饅頭

製作步驟 / Step

1 用8毫升温水泡發4克燕麥片。將泡發的燕麥片、麵粉、糖、酵母、泡打粉、20毫升水及葵花籽油混合在一起揉成麵團，發酵1小時。

Soak 4g of the oatmeal in 8ml of warm water. Mix the soaked oatmeal, flour, sugar, yeast, baking powder, 20ml water and sunflower oil together and let the dough rise for 1 hour.

2 將麵團分份，揉成饅頭形狀，二次醒發10分鐘。蒸鍋上汽，放入饅頭蒸15分鐘，關火後靜置5分鐘，開蓋即可。

Divide the dough and shape into buns. Let the buns rise for another 10 minutes. Steam for 15 minutes, then turn off the heat and let rest for 5 minutes before serving.

優點及營養成份

一份燕麥全麥饅頭它既注重口感，又強調營養均衡，與普通白麵饅頭相比，增加了奇亞籽燕麥片會使饅頭更有嚼勁，是一個不錯的主食選擇。

燕麥洋蔥雞肉卷

早餐系列
Breakfast

食材份量 / Ingredient

火腿 15g/克
Ham

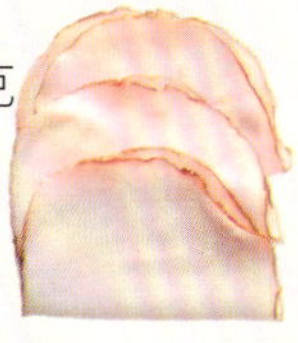

全麥捲餅皮 1pc/片
Whole Wheat Wrap

馬蘇里拉芝士 10g/克
Mozzarella Cheese

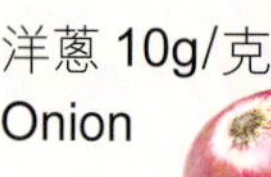

洋蔥 10g/克
Onion

蛋黃醬 5g/克
Mayonnaise

生菜（適量）
Lettuce a pinch

雞胸肉 20g/克
Chicken Breast

桂格快熟燕麥片 30g/克
Quick Cook Oatmeal

熱水 200ml/毫升
Hot Water

Oat and Onion Chicken Roll
燕麥洋蔥雞肉卷

製作步驟 / Step

1 雞肉、洋蔥炒香。燕麥片用熱開水泡1分鐘後撈出瀝乾待用。
Soak the oatmeal in hot water for 1 minute. Stir fried chicken and onion and set aside.

2 熱鍋，放入全麥捲餅皮，小火煎熟；將泡好的燕麥片、火腿、洋蔥雞肉、生菜、蛋黃醬及馬蘇里拉芝士，放在薄餅上，捲起。
Warm the whole wheat wrap with low heat. Place the soaked oatmeal, ham, onion, chicken, lettuce, mayonnaise and mozzarella cheese on the wrap before wrapping.

優點及營養成份

與市售傳統雞肉卷相比，使用雜糧麵餅皮和快煮燕麥，有助於增強飽腹感，平穩血糖。雞肉和馬蘇里拉芝士是優質蛋白的良好來源。一份燕麥洋蔥雞肉卷提供的熱量適中，再搭配一杯牛奶，是營養健康早餐的不錯選擇。

燕麥香蕉吐司

食材份量 / Ingredient

早餐系列
Breakfast

混合堅果 20g/克
Mix Nuts

香蕉 1pc/隻
Banana

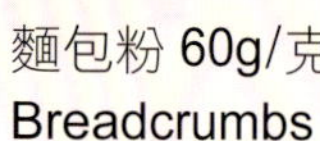

麵包粉 60g/克
Breadcrumbs

酵母 3g/克
Yeast

麵包改良劑 1g/克
Bread Improver

桂格即食燕麥片 40g/克
Instant Oatmeal

牛油 4g/克
Butter

鹽 1.5g/克
Salt

溫水 75ml/毫升
Water

Banana Oat Toast
燕麥香蕉吐司

1 所有原料混合拌勻成團，放置20分鐘後搓成團，表面撒上燕麥片。

Mix all the ingredients to form a dough and let it rest for 20 minutes. Shape the dough, then sprinkle the oatmeal over the surface.

2 團放入烤盤，在溫度35°C-40°C下擺放1小時發烤。放入烤箱用220°C烘烤20-25分鐘。

Let the dough rise at 35°C-40°C for 1 hour, then bake at 220°C for 20-25 minutes.

優點及營養成份

相比於普通吐司，燕麥香蕉吐司具有控制總體熱量攝入的同時，它還能提供持久的飽腹感，適合健身、體重控制的人。

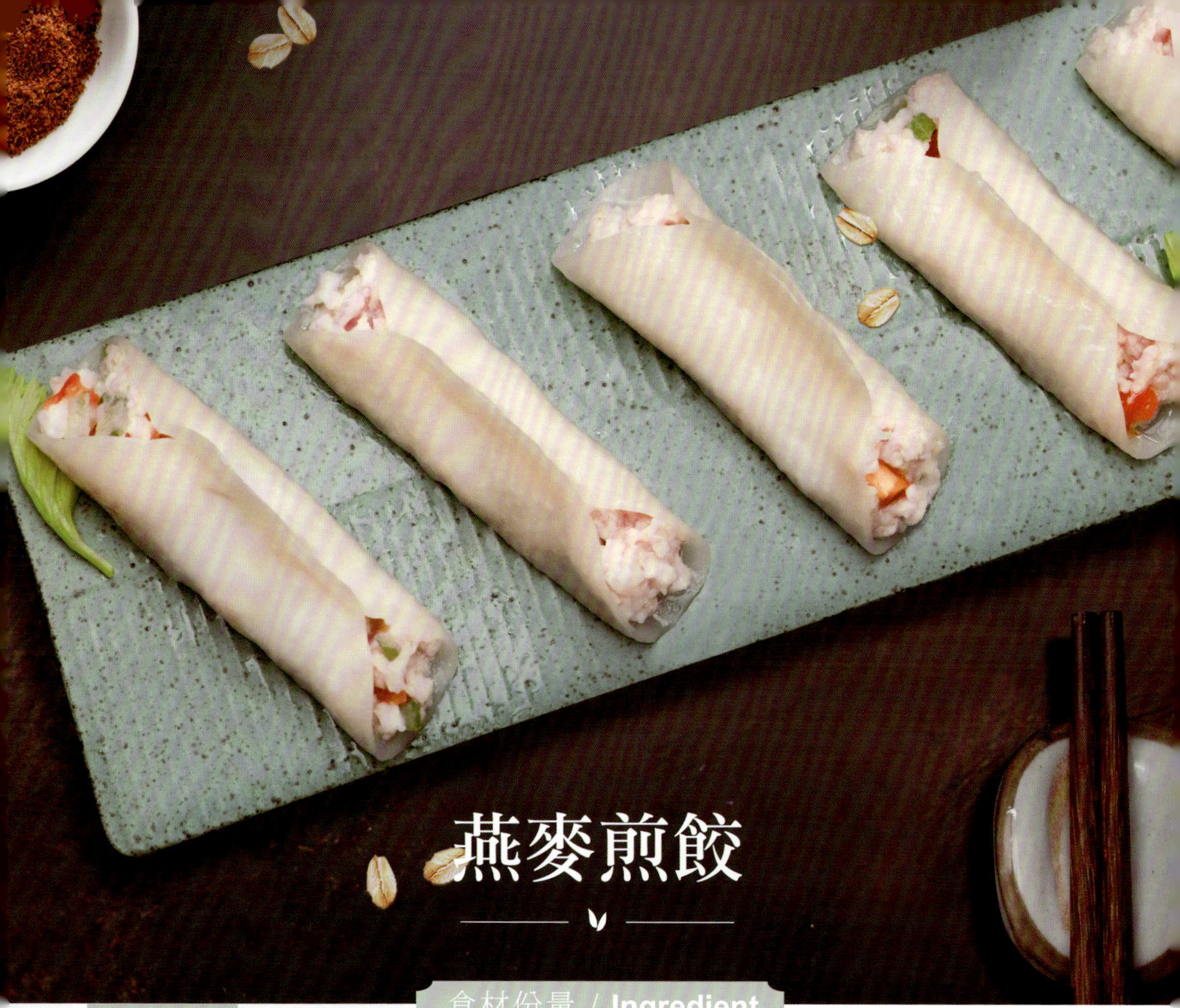

燕麥煎餃

食材份量 / Ingredient

早餐系列
Breakfast

馬蹄（荸薺）25g/克
Chinese Water Chestnut

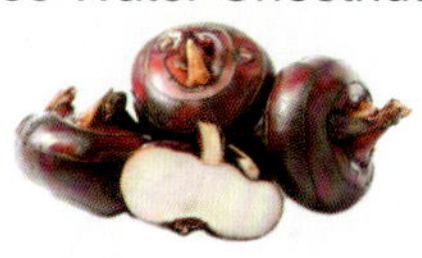

餃子皮 80g/克
Dumpling Skin

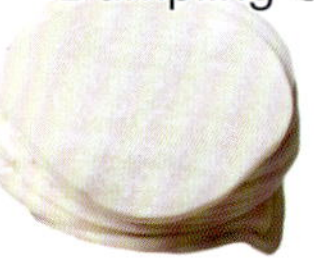

芹菜 25g/克
Celery

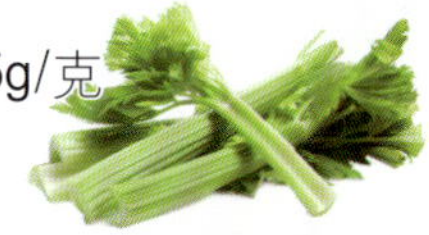

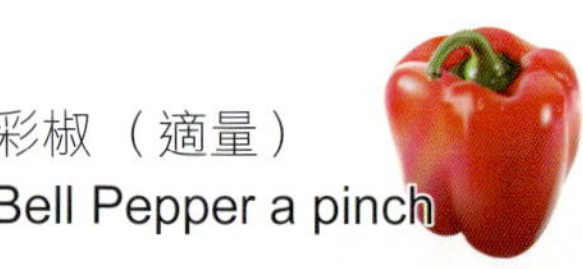

彩椒 （適量）
Bell Pepper a pinch

豬肉 125g/克
Pork

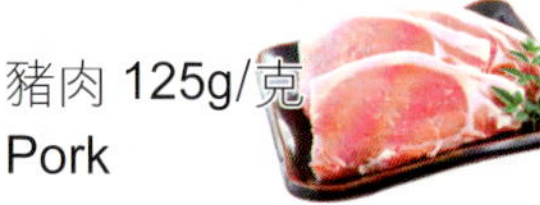

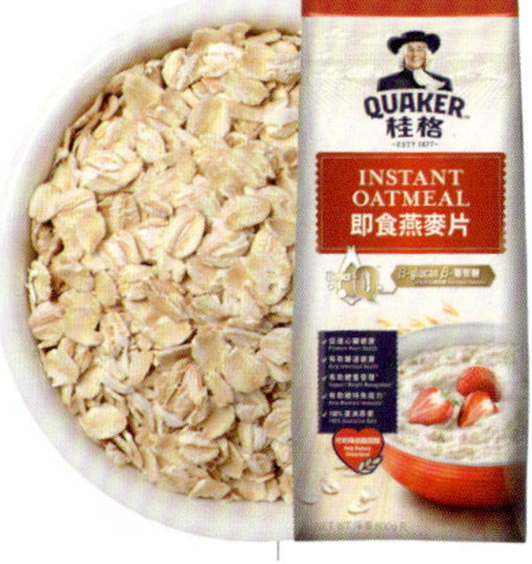

桂格即食燕麥片 20g/克
Instant Oatmeal

洋蔥 10g/克
Onion

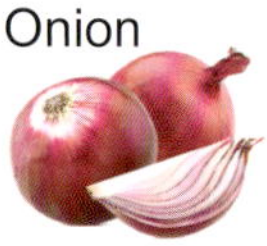

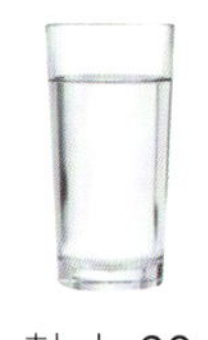

熱水 60ml/毫升
Hot Water

調味料 / Seasoning

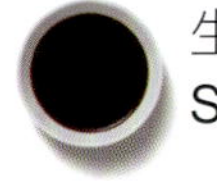

生抽 1ml/毫升
Soy Sauce

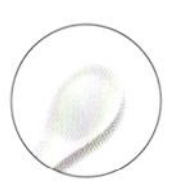

鹽 1.5g/克
Salt

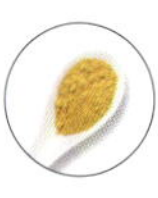

胡椒粉 1g/克
Pepper

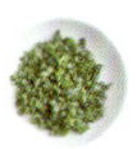

蔥花 5g/克
Scallion

Oat Fried Dumpling
燕麥煎餃

製作步驟 / Step

1 將燕麥片用熱水泡發5分鐘，待用。
Soak the oatmeal in hot water for 5 minutes and set aside.

2 將洋蔥、芹菜、彩椒、馬蹄切粒，拌入豬肉糜和泡好的燕麥片，加入胡椒粉、鹽、醬油調味拌勻。
Mix the soaked oatmeal with ground pork, chopped onion, celery, bell pepper and Chinese water chestnut. Season with pepper, salt and soy sauce.

3 將拌勻的燕麥豬肉餡包入餃子皮，熟鍋放油，放入餃子，小火煎熟，起鍋撒蔥花。
Use mixture as dumpling as filling for the dumpling skins, then fry it over low heat. Sprinkle with chopped scallion before serving.

優點及營養成份

將燕麥加入煎餃餡料中，不僅提升了口感，還增加了膳食纖維含量，有益於胃腸道健康。這道菜融合了多種健康食材，如肉類、穀類和蔬菜類，色彩鮮亮、營養豐富。

燕麥素包

早餐系列
Breakfast

食材份量 / Ingredient

青菜 1pc/顆
Vegetable

豆腐乾 1pc/件
Dried Tofu

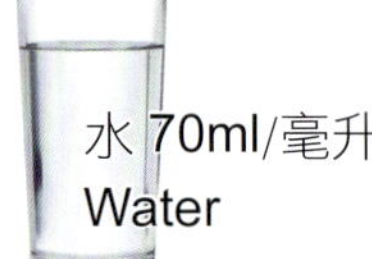
水 70ml/毫升
Water

筍 6g/克
Bamboo Shoots

葵花籽油 5ml/毫升
Sunflower Oil

桂格5黑混合即食麥片 25g/克
5 Black Multi-Grain

麵粉 65g/克
Flour

酵母 1g/克
Yeast

泡打粉 1g/克
Baking Powder

調味料 / Seasoning

鹽 1g/克
Salt

砂糖 2.5g/
Sugar

Veggie Oat Bun
燕麥素包

製作步驟 / Step

1 20克燕麥片加40毫升熱水泡發，泡發後加麵粉、泡打粉、酵母、糖、30毫升温水拌勻成團，分成小麵團。
Soak 20g of oatmeal in 40ml warm water. Mix the soaked oatmeal with flour, baking powder, yeast, sugar and 30ml warm water. Divide the dough into small portions.

2 青菜、筍、豆腐乾分別燙熟切碎拌勻，加入葵花籽油及所有調味料，包入燕麥麵團。
Chop the vegetables, bamboo shoots and dried tofu. Mix with sunflower oil, salt and sugar. Wrap the filling in the oatmeal dough portions.

3 放蒸籠，在溫度20°C-25°C發烤1小時。蒸烤箱100°C蒸6-10分鐘。
Let the buns rise at 20°C-25°C for 1 hour. Steam at 100°C for 6-10 minutes.

優點及營養成份

青菜、筍、豆腐乾等蔬菜使包子餡料的口感豐富多樣，同時可提供豐富的植物性蛋白、維他命 A、維他命 B 雜、鈣、鐵和膳食纖維，適合素食者或者需要減少肉類攝入的人食用。

五福燕麥燒賣

早餐系列 *Breakfast*

食材份量 / Ingredient

燒賣皮 30g/克
Dumpling Skin

蝦仁 4pcs/隻
Prawns

萵筍 40g/克
Celtuce

胡蘿蔔 40g/克
Carrot

香菇 2pcs/隻
Mushroom

雞胸肉 120g/克
Chicken Breast

洋蔥 20g/克
Onion

桂格即食燕麥片 40g/克
Instant Oatmeal

糯米 220g/克
Sticky Rice

熱水 120ml/毫升
Hot Water

調味料 / Seasoning

酒 1ml/毫升
Cooking Wi

生抽 10ml/毫
Soy Sauce

鹽 1.5g/克
Salt

胡椒粉 1g/克
Pepper

Wufu Oat Siu Mai
五福燕麥燒賣

製作步驟 / Step

1 糯米加水，電飯煲蒸煮30分鐘，煮熟冷卻待用；燕麥片熱水泡發5分鐘，待用。
Cook sticky rice in the rice cooker for 30 minutes and let it cool. Soak the oatmeal in hot water for 5 minutes and set aside.

2 雞肉和所有蔬菜（萵筍、香菇、洋蔥、胡蘿蔔）切粒炒熟，加糯米飯和燕麥片，醬油調味；蝦仁中放醬油、料酒、胡椒粉、鹽進行醃製。
Chop chicken breast and all vegetables (celtuce, mushroom, onion, carrot). Mix with sticky rice and oatmeal, then add soy sauce and mix well. Marinate the prawns with soy sauce, cooking wine, pepper and salt.

3 在燒麥皮中放入燕麥糯米飯餡料，包起，在面放上蝦仁，放入蒸鍋，水開蒸8-10分鐘。
Place the oatmeal and sticky rice mixture onto dumpling skin with prawns. Steam over boiling water for 8-10 minutes.

優點及營養成份

五福燕麥燒麥採用多種蔬菜粒和肉粒，既提供了優質蛋白質、維他命、礦物質等營養物質，也增添了豐富的色彩和清爽的口感，燕麥片具有高膳食纖維和低 GI 的特點，對於健康飲食具有積極作用。

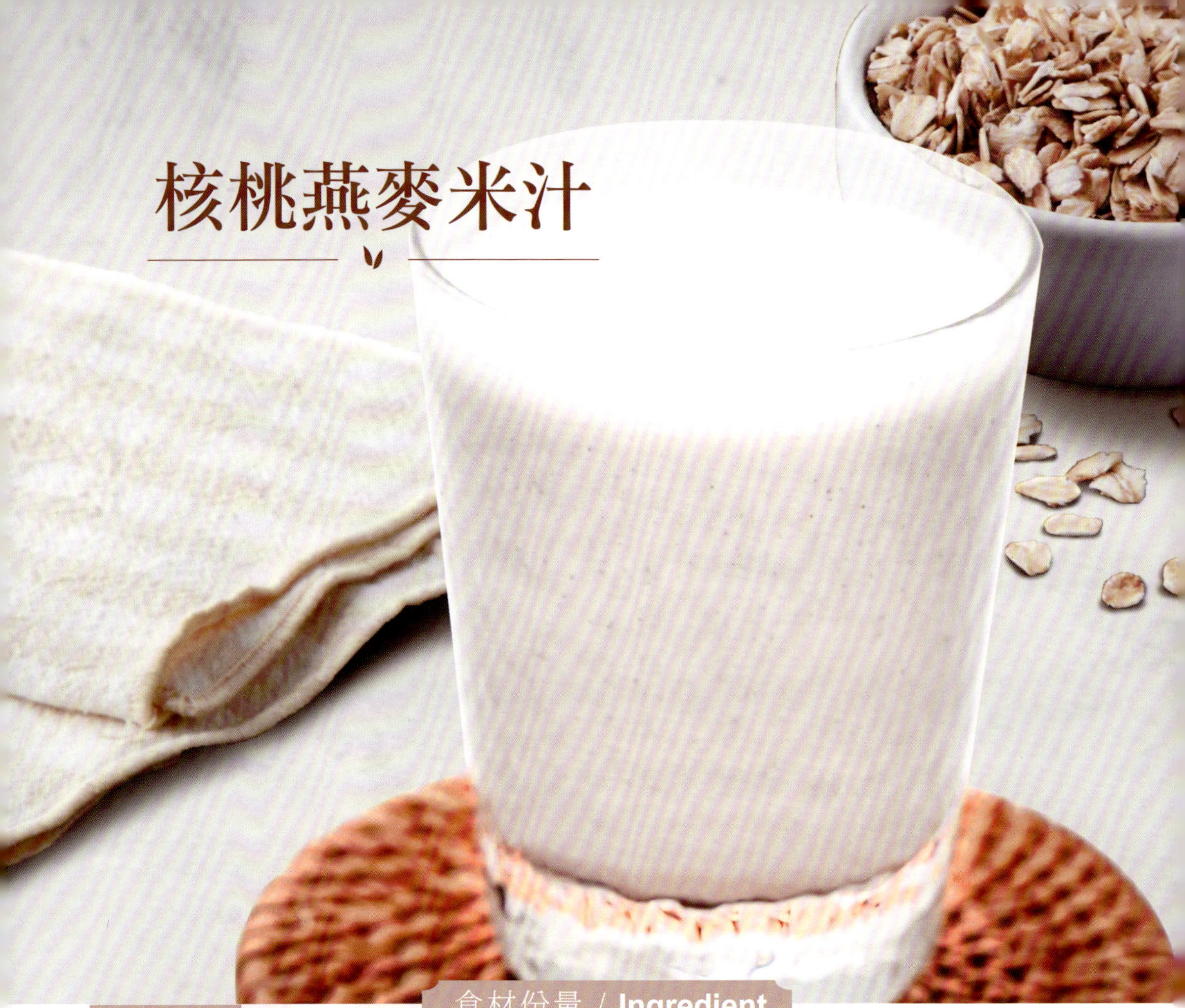

核桃燕麥米汁

早餐系列
Breakfast

食材份量 / Ingredient

冰糖 5g/克
Rock Sugar

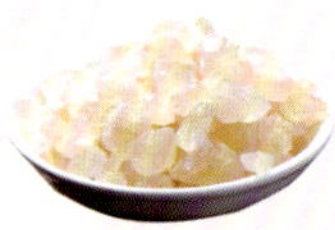

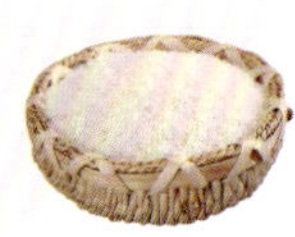

大米 10g/克
Rice

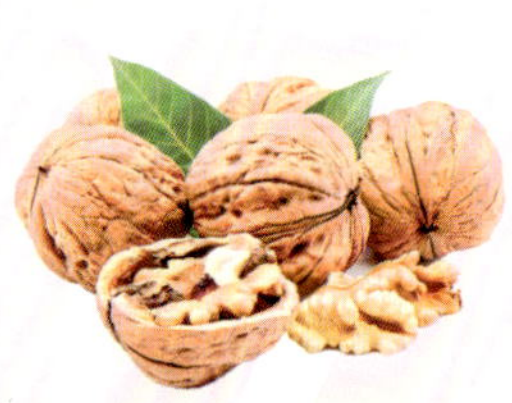

核桃 2pcs/顆
Walnut

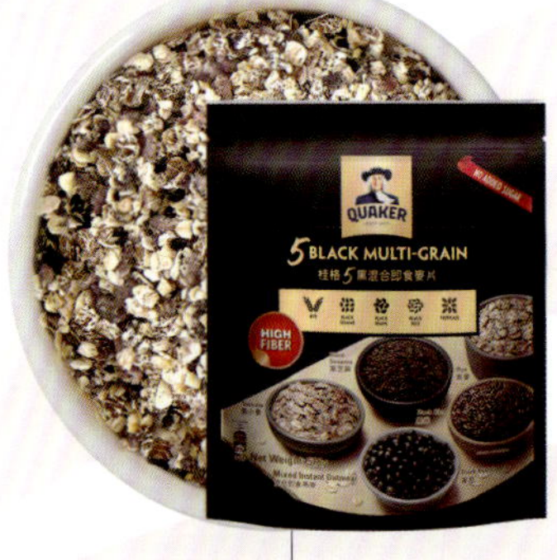

桂格5黑混合即食麥片 10g/克
5 Black Multi-Grain

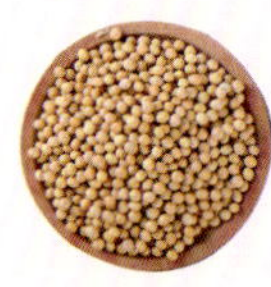

黃豆10g/克
Soya Bean

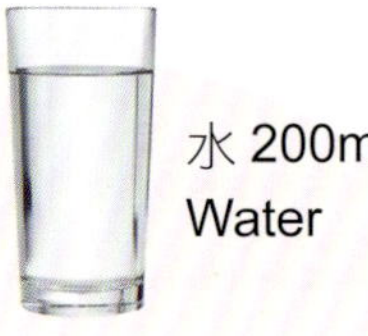

水 200ml/毫升
Water

Walnut Oatmeal Rice Milk
核桃燕麥米汁

製作步驟 / Step

1 提前將黃豆浸泡8小時以上，大米浸泡2小時；將泡好的黃豆、大米、核桃、燕麥片和適量清水一起加入豆漿機（或破壁機），啟動豆漿功能。
Soak soya beans and rice for 2 hours separately. Add the soaked soya beans, rice, walnuts, oatmeal, and water into a soya bean milk grinder. Blend until smooth.

2 打好的米汁加入冰糖調勻即可飲用。
Add rock sugar to taste before serving.

優點及營養成份

核桃燕麥米汁由大米、黃豆、核桃仁以及 5 黑混合即食燕麥片混合而成，形成了一款豐富多元的飲品。這款飲品製作步驟簡單，適合現代快節奏的生活，可作為早餐或飯後飲用。

五紅補氣燕麥豆漿

早餐系列
Breakfast

食材份量 / Ingredient

枸杞 5g/克
Goji Berry

紅豆 15g/克
Red Bean

薏仁（薏米）5g/克
Job's Tears

熱豆漿 240ml/毫升
Hot Soya Milk

桂格5紅混合即食麥片 30g/克
5 Red Multi-Grain

桂圓肉 15g/克
Dried Longan

Five Reds Nourishing Oat Soy Milk
五紅補氣燕麥豆漿

1 將蒸煮好的薏仁、紅豆、去核桂圓肉、枸杞放入杯中。
Place the steamed job's tears, red bean, dried longan pulp, goji berry in a mug.

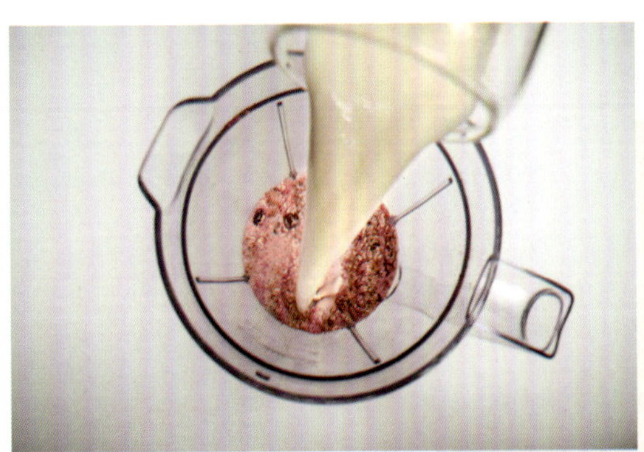

2 將熱豆漿倒入燕麥片中泡發5分鐘，放入攪拌機攪打均勻。
Pour hot soya milk into the mug with the oatmeal and soak for 5 minutes, then blend the mixture until smooth.

3 將燕麥豆漿倒入杯中即製作完成。
Pour the blended oatmeal soya milk into the mug before serving.

優點及營養成份

這款豆漿使用桂圓、紅豆的天然甜味代替白糖，可以有效減少精製糖的攝入。它是一道營養豐富、口感香甜、顏色誘人的飲品。

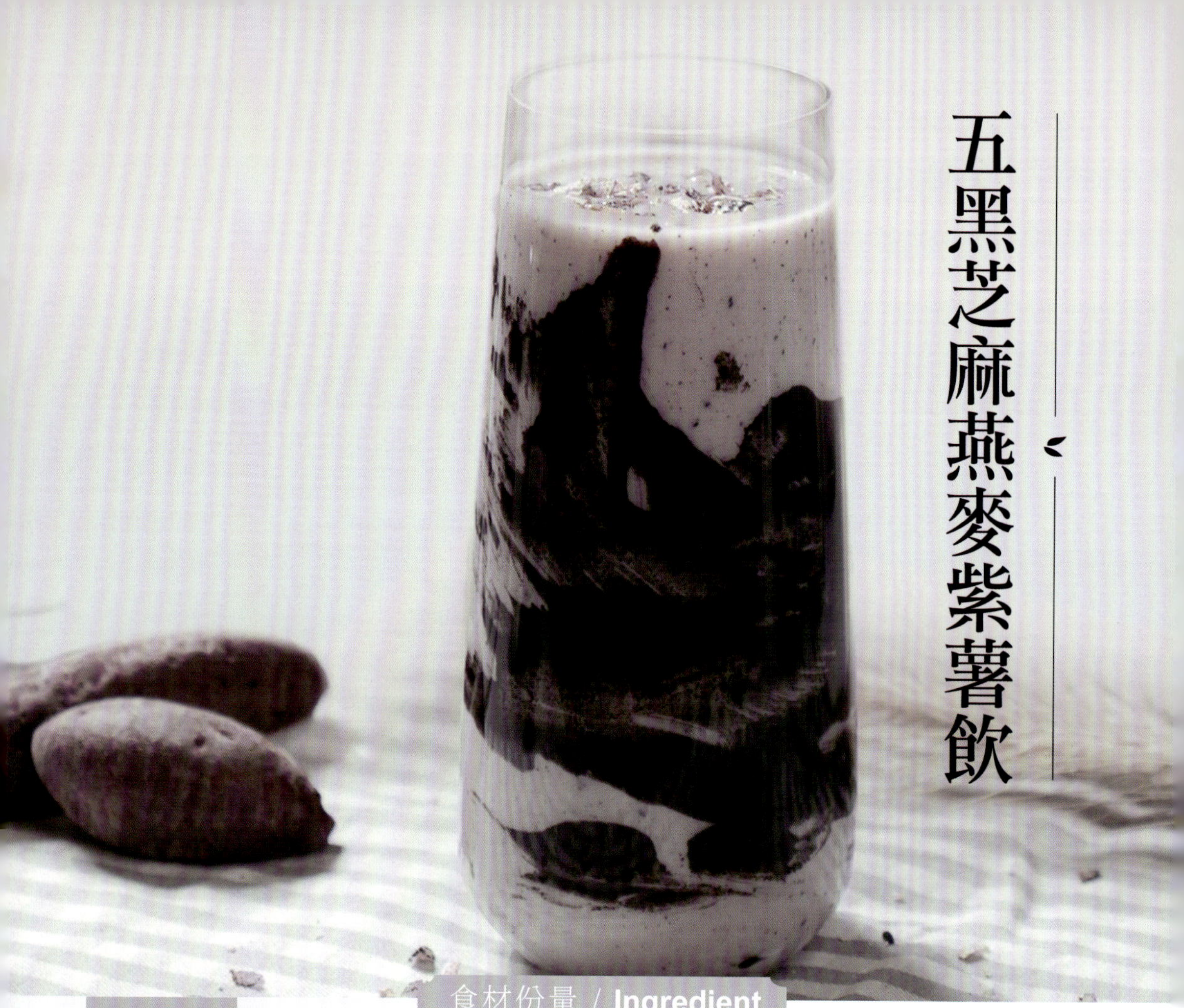

五黑芝麻燕麥紫薯飲

食材份量 / Ingredient

黑芝麻醬 10g/克
Black Sesame Paste

紫薯（熟）20g/克
Cooked Purple Sweet Potato

桂格5黑混合即食麥片 30g/克
5 Black Multi-Grain

牛奶
Milk 150ml/毫升

Five Black, Sesame, and Purple Sweet Potato Oat Drink

五黑芝麻燕麥紫薯飲

製作步驟 / Step

1 將芝麻醬貼杯放入杯中，加入蒸煮過的紫薯粒。

Place the cooked purple sweet potato cubes, black sesame and oatmeal in a mug.

2 將熱牛奶倒入燕麥片中泡發5分鐘，用攪拌機攪打均勻。

Pour hot milk into the mug and soak for 5 minutes, then blend the mixture until smooth.

3 將燕麥乳倒入放有紫薯的燕麥杯中即可。

Pour the blended oatmeal and purple sweet potato drink into a mug before serving.

優點及營養成份

這款飲料使用了桂格 5 黑混合即食燕麥片作為主要食材，為飲品提供了多樣性的營養素。紫薯粒的綿密口感再搭配上溫暖的熱牛奶，既滿足了口味的多樣性，又提升了綜合營養價值。無論是老人、小孩還是上班族，都可以在早餐中享用一杯。

奶香蘑菇配燕麥燴水波蛋

早餐系列 *Breakfast*

食材份量 / Ingredient

蘑菇 100g/克
Mushroom

雞蛋 2pcs/隻
Egg

牛奶
Milk 320ml/毫升

雞胸肉 50g/克
Chicken Breast

桂格即食燕麥片 80g/克
Instant Oatmeal

鹽 1.5g/克
Salt

Creamy Mushroom With Oat and Poached Egg
奶香蘑菇配燕麥燴水波蛋

製作步驟 / Step

1 起鍋熱油，放入雞胸肉、蘑菇一起炒香，炒香後鍋中加入牛奶，煮開後放入燕麥片。
Spray oil over pan, fry the chicken breast and mushroom until cooked thoroughly . Add the milk and bring to a boil, then add the oatmeal.

2 調味後盛出，放上水波蛋。
Season the oatmeal mixture with salt, then place the poached egg on the top before serving.

優點及營養成份

燕麥、蘑菇和牛奶的混合鮮香四溢，口感濃郁。低脂牛奶代替芝士更低卡，保持奶香口感的同時保證了蛋白質的供應。蘑菇膳食纖維豐富，並且含有多種微量元素。整體而言，這是一道兼具創新和健康的美食。

燕麥厚蛋燒

食材份量 / **Ingredient**

早餐系列 *Breakfast*

青瓜 15g/克
Cucumber

食油 2ml/毫升
Cooking Oil

雞蛋 1pc/隻
Egg

鹽 0.2g/克
Salt

胡蘿蔔 15g/克
Carrot

水 60ml/毫升
Water

桂格即食燕麥片 20g/克
Instant Oatmeal

Oatmeal Omelet
燕麥厚蛋燒

製作步驟 / Step

1 先將燕麥片用適量開水浸泡5分鐘，將胡蘿蔔、青瓜切成小碎粒備用。
Soak in the oatmeal with water for 5 minutes, then chop carrot and cucumber into small cubes for later use.

2 泡軟的燕麥片放入60毫升清水攪拌成糊狀，加入雞蛋、胡蘿蔔、黃瓜，再加入鹽，繼續攪拌均勻。
Add 60ml water into the soaked oatmeal and mix it into semi-liquid form, then mix it well with egg, carrot, cucumber and salt.

3 熱鍋下油，用小火煎燕麥蛋液，均勻平鋪在鍋底至半凝狀從右向左卷起，然後推向右邊，往左邊繼續加入燕麥蛋液 ，重複以上步驟。
Heat oil in a pan over low heat. Pour in oatmeal egg mixture and spread evenly. When the mixture is half set, fold one side over, then pour more mixture onto the left side. Repeat the above step.

4 複至用完蛋液，出鍋切塊。
Cut into small portions before serving.

優點及營養成份

燕麥厚蛋燒製作簡單，在家裏能夠輕鬆完成。燕麥和雞蛋一起混合提高了燕麥中蛋白質的利用率，胡蘿蔔和黃瓜增加了菜品的色彩感，提高用餐人的食慾。這款燕麥厚蛋燒食譜巧妙地結合了營養和美味，是一道既能當早餐，又能當點心的菜餚。

西蘭花燕麥烘蛋

早餐系列
Breakfast

食材份量 / Ingredient

西蘭花 140g/克
Broccoli

雞蛋 1pc/隻
Egg

鹽 0.5g/克
Salt

黑胡椒 0.5g/克
Black Pepper

櫻桃番茄 7pcs/顆
Cherry Tomato

桂格5紅混合即食麥片 50g/克
5 Red Multi-Grain

牛奶 100ml/毫升
Milk

Baked Egg With Broccoli and Oats
西蘭花燕麥烘蛋

製作步驟 / Step

1 盤中放入燕麥片，倒上牛奶浸泡3分鐘。
Soak the oatmeal in milk for 3 minutes.

2 西蘭花洗乾淨切小朵，用熱水燙一下，櫻桃番茄對半切開。
Clean and boil chopped broccoli. Cut cherry tomatoes in half.

3 在烤盤中鋪上一層泡發的燕麥，放入西蘭花、櫻桃番茄 。
Place a layer of the soaked oatmeal in a baking dish, then add broccoli and cherry tomatoes.

4 雞蛋打成蛋液，用鹽和黑胡椒調味，倒入烤盤中。
Pour the beaten egg and season with salt and pepper.

5 烤箱180°C，20分鐘即可。
Bake at 180°C for 20 minutes.

優點及營養成份

這道菜的特色在於將燕麥與雞蛋相結合，形成口感豐富的半凝狀烘蛋，展現了獨特的風味。而且製作簡單，營養豐富，可作為早餐的快手菜品選擇之一。

生椰拿鐵燕麥

早餐系列
Breakfast

食材份量 / Ingredient

芒果 30g/克
Mango

香蕉 30g/克
Banana

椰奶 100ml/毫升
Coconut Milk

濃縮咖啡 40ml/毫升
Espresso

優酪乳（乳酪）
Yogurt 50g/克

桂格快煮燕麥片 35g/克
Quick Cook Oatmeal

薄荷葉 1g/克
Mint Leaves

Coconut Latte With Oats
生椰拿鐵燕麥

製作步驟 / Step

1 杯中放入香蕉，然後放入燕麥片，將椰奶和濃縮咖啡倒入杯中，放入冰箱冷藏一晚。
Put banana, quick cooked oatmeal, coconut milk and espresso in a mug. Mix well and place in the fridge overnight.

2 早上將放了冷藏一夜的燕麥取出，再倒入優酪乳。
Take the oats out of the fridge and add yogurt next morning.

3 放入芒果，薄荷葉裝飾。
Garnish with mango and mint leaves before serving.

優點及營養成份

經過浸泡的燕麥變得更加軟糯，口感更接近蛋糕。冷藏一夜後的燕麥產生了更多的抗澱粉性，比普通燕麥吸收更少的熱量，同時還能維持飽腹感。這款飲料不僅營養豐富，而且味道獨特，是一款值得一試的美味選擇。

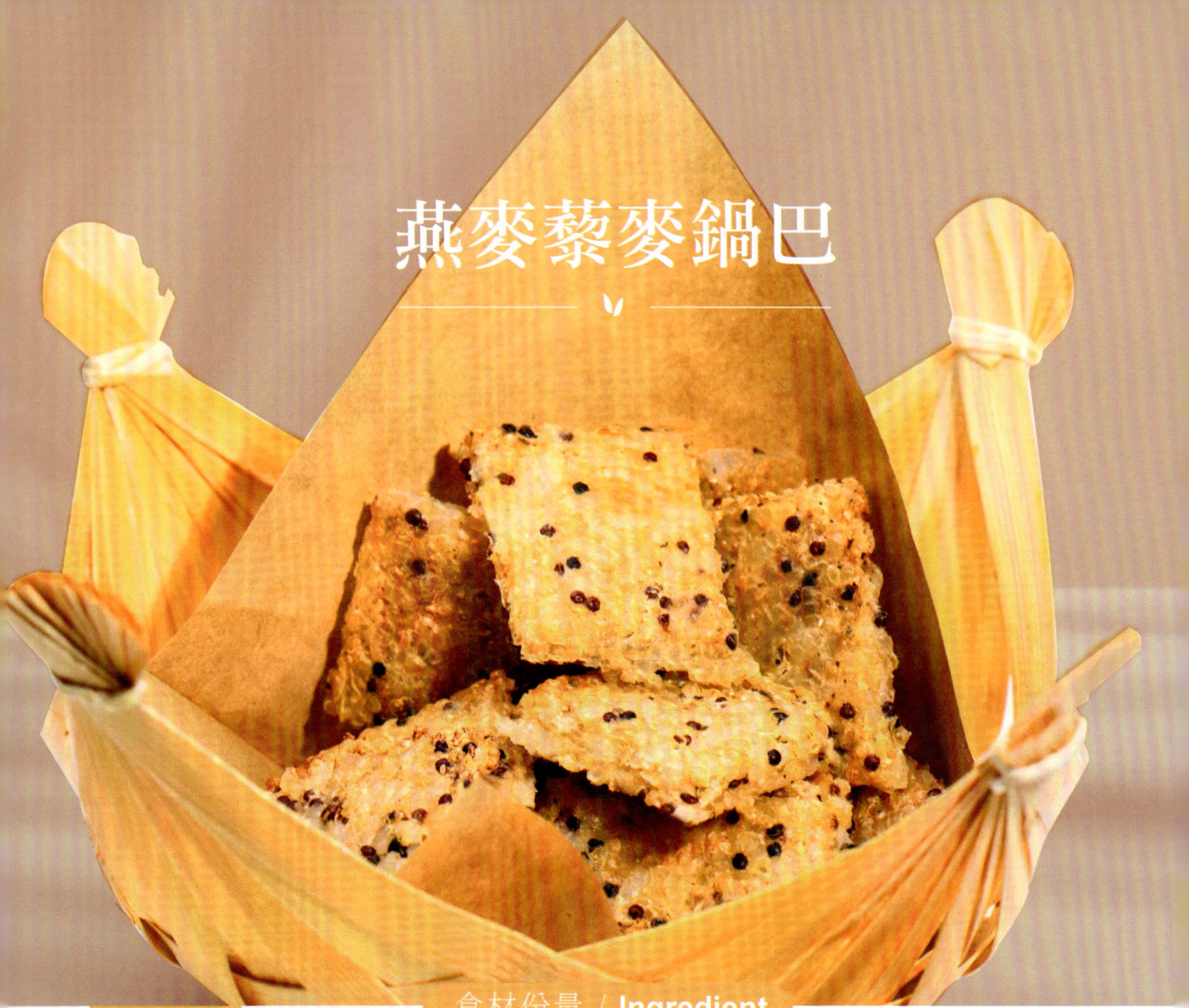

燕麥藜麥鍋巴

食材份量 / **Ingredient**

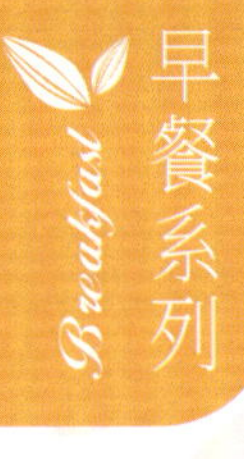

早餐系列

Breakfast

大米 50g/克
Rice

黑胡椒 1g/克
Black Pepper

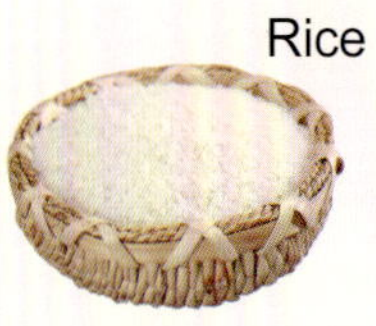

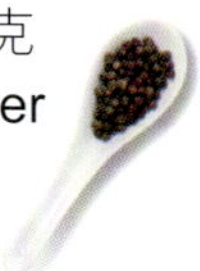

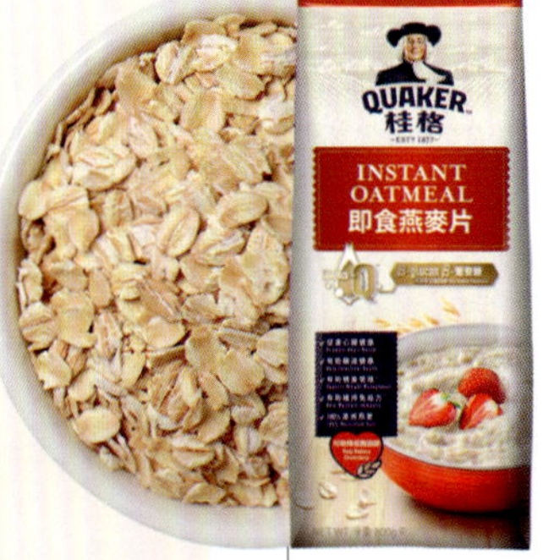

桂格即食燕麥片 30g/克
Instant Oatmeal

鹽 0.3g/克
Salt

藜麥 20g/克
Quinoa

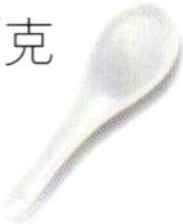

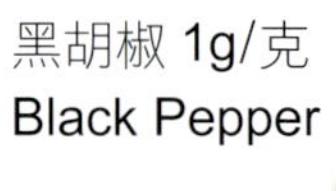

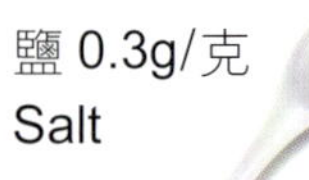

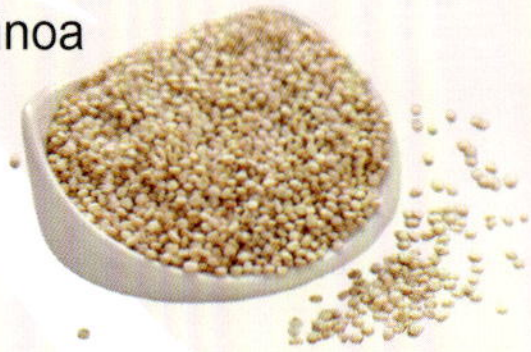

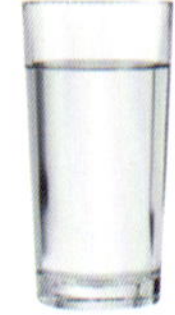

水 110ml/毫升
Water

Oat and Quinoa Crispy Rice
燕麥藜麥鍋巴

製作步驟 / Step

1 燕麥片、藜麥和白米蒸成米飯後放涼備用。
Cook the oatmeal, quinoa and rice and set aside for later use.

2 將燕麥米飯放在烤盤中攤平、整形。
Place the cooked mixture in baking pan. Spread and shape evenly.

3 切小塊後放入烤箱，將鹽、黑胡椒均勻灑在表面。
Divide into small portions and sprinkle salt and black pepper evenly on surface for seasoning.

4 烤箱預熱，放入中層，180°C烘烤20分鐘即可。
Pre-heat the oven, bake at 180°C for 20 minutes at middle level.

優點及營養成份

燕麥藜麥鍋巴替換了部分精製大米，添加了燕麥片和藜麥米，這兩種雜糧含有維他命 B 雜和膳食纖維，幫助控制飲食 GI。燕麥藜麥鍋巴口感酥脆，製作過程簡便，不需要複雜的廚房技巧，為廚房新手提供了嘗試新食譜的機會。

燕麥西蘭花薯餅

早餐系列
Breakfast

食材份量 / Ingredient

西蘭花 140g/克
Broccoli

馬鈴薯 300g/克
Potato

胡蘿蔔 70g/克
Carrot

雞蛋 1pc/隻
Egg

桂格即食燕麥片 50g/克
Instant Oatmeal

芝士碎 60g/克
Grated Cheese

芝士粉 15g/克
Cheese Powder

黑胡椒 0.5g/克
Black Pepper

鹽 1g/克
Salt

Oat and Broccoli Hash Browns
燕麥西蘭花薯餅

製作步驟 / Step

1 馬鈴薯去皮切成厚片，上蒸鍋大火蒸15分鐘左右熟透，用叉子壓碎成薯泥。
Remove the skin and chop potatoes into pieces. Steam for 15 minutes and mash it to puree.

2 西蘭花、胡蘿蔔切小粒。
Chop broccoli and carrot into small pieces.

3 薯泥中加入西蘭花、胡蘿蔔、雞蛋、燕麥片、芝士碎、芝士粉、鹽和黑胡椒碎，拌勻成團，外面沾上燕麥片。
Mix the broccoli, carrot, egg, oatmeal, grated cheese, cheese powder, salt and black pepper into the mashed potatoes and cover it with oatmeal.

4 薯餅放入烤箱以200°C焗烤10分鐘後取出，翻轉面繼續烤10分鐘左右，表面金黃即可。
Bake at 200°C for 10 minutes, then flip the hash browns and bake for another 10 minutes until both sides are golden brown.

優點及營養成份

馬鈴薯可以提供能量和膳食纖維，西蘭花和胡蘿蔔含維他命、礦物質和抗氧化劑，燕麥片增加了全穀物纖維和β-葡聚醣，芝士則注入了鈣元素和優質蛋白質。此外，採用烤的方式減少了食物的油脂攝入，使得這款美味的薯餅更適合作為控制飲食的好選擇。

香蕉燕麥烤布丁

早餐系列
Breakfast

食材份量 / Ingredient

雞蛋 1pc/隻
Egg

香蕉 200g/克
Banana

牛奶 100ml/毫升
Milk

桂格即食燕麥片 50g/克
Instant Oatmeal

Baked Banana Oat Pudding
香蕉燕麥烤布丁

製作步驟 / Step

1 香蕉壓成泥打入雞蛋，倒入燕麥片和牛奶混合攪拌均勻，倒入焗碗中。
Mix banana puree with egg, Add the oatmeal and milk and mix well. Pour the mixture into a baking mug.

2 另半根香蕉切成片，將香蕉片放在燕麥牛奶糊上，烤箱提前預熱，180°C烤20-25分鐘。
Slice another banana and place the slices on the top of the oatmeal mixture. Pre-heat the oven and bake at 180°C for 20-25 minutes.

優點及營養成份

香蕉燕麥烤布丁香甜軟糯、飽腹，香蕉泥、雞蛋、燕麥和牛奶混合，展現了簡單而有效的食材搭配，能起到營養互補的作用，使機體更易吸收。所有食材攪打成泥後，只需放入烤箱就能完成布丁的製作，方便簡單。

桂格5紅混合即食燕麥片

間餐系列

Snacks

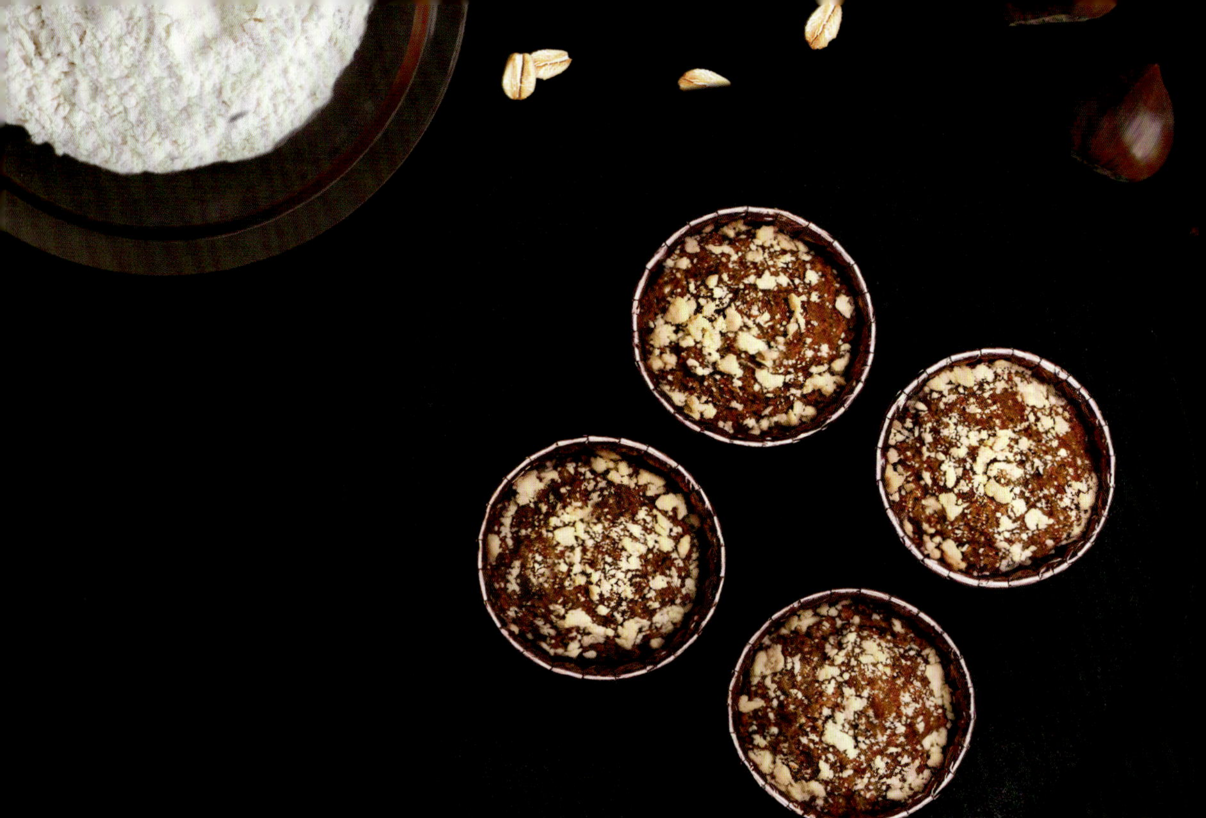

栗蓉燕麥鬆餅

間餐系列 *Snacks*

食材份量 / Ingredient

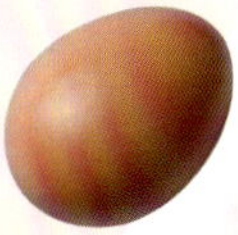

雞蛋 1pc/隻
Egg

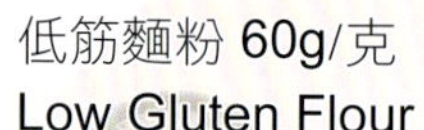

低筋麵粉 60g/克
Low Gluten Flour

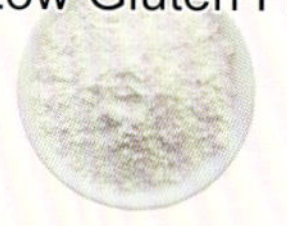

泡打粉 2g/克
Baking Powder

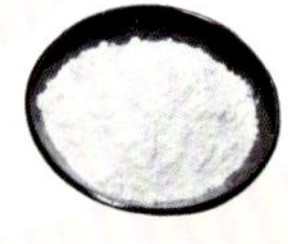

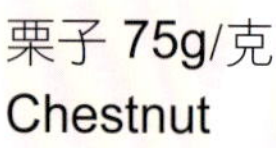

栗子 75g/克
Chestnut

桂格即食燕麥片 40g/克
Instant Oatmeal

植物油 50g/克
Vegetable Oil

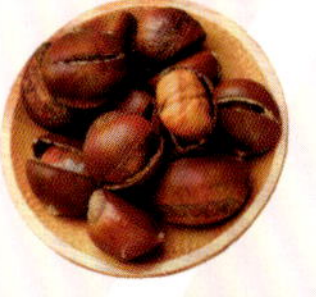

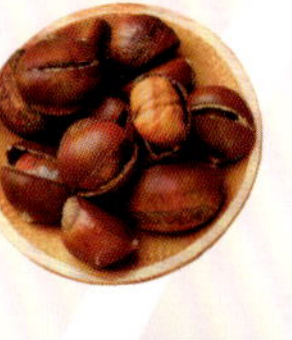

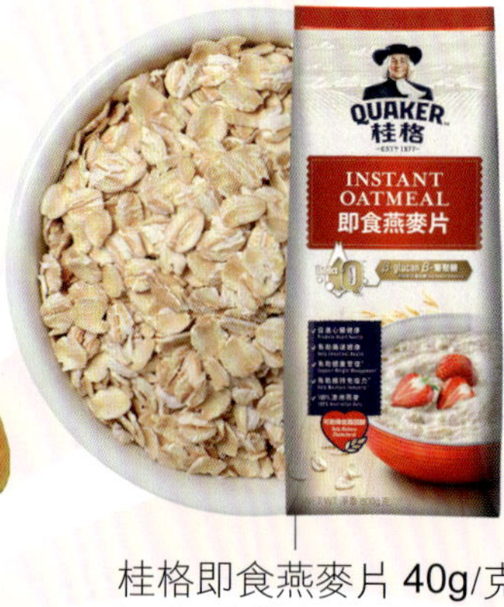

南瓜泥 50g/克
Pumpkin Puree

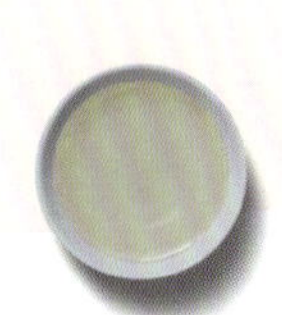

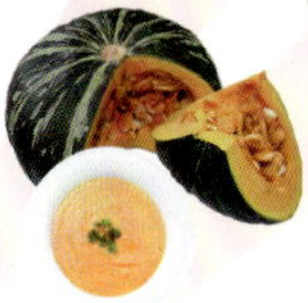

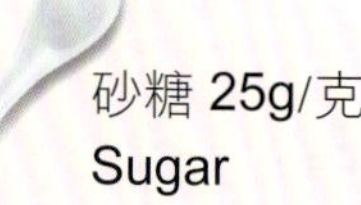

砂糖 25g/克
Sugar

Chestnut Oat Muffins
栗蓉燕麥鬆餅

製作步驟 / Step

1 南瓜去皮切成小塊，用蒸約15分鐘，壓制成泥。將熟栗子壓碎後倒入南瓜泥中。
Remove the pumpkin skin, chop into pieces, and steam for 15 minutes. Mash into a puree. Add cooked chestnuts and mash together with the pumpkin puree.

2 將雞蛋、砂糖、植物油、南瓜泥和燕麥混合打勻。
Mix the egg, sugar, vegetable oil, oatmeal, and the pumpkin, chestnut puree until well combined.

3 將低筋麵粉、泡打粉、燕麥片加入蛋糕液體，攪拌混合成麵糊。麵糊倒入紙杯，撒燕麥片點綴即可。
Add the low gluten flour, baking powder and oatmeal into the puree and mix it into a dough. Portion the dough into paper cup and sprinkle with oatmeal.

4 預熱烤箱，175°C烤20分鐘。用牙籤叉進蛋糕，拔出沒有附著麵糊即代表烤熟。
Pre-heat the oven, bake at 175°C for 20 minutes. A fully cooked cake will have no sticky batter on a toothpick inserted into the center.

優點及營養成份

栗蓉燕麥鬆餅以多種營養豐富的食材為基礎，呈現出獨特的口感和風味。使用燕麥和新鮮栗子及南瓜降低了精製糖的攝入，符合飲食控制與健康的需求。

紅薯燕麥甜甜圈

間餐系列 *Snacks*

食材份量 / Ingredient

紅薯 100g/克
Sweet Potato

混合堅果 10g/克
Mixed Nuts

蛋黃 17g/克
Egg Yolk

桂格即食燕麥片 20g/克
Instant Oatmeal

Sweet Potato Oat Donuts
紅薯燕麥甜甜圈

製作步驟 / Step

1 紅薯切片，上鍋蒸熟，趁熱搗成紅薯泥。加入燕麥片，攪拌均勻。
Slice sweet potato into pieces, steam until soft, and mash it to puree. Mix in the oatmeal.

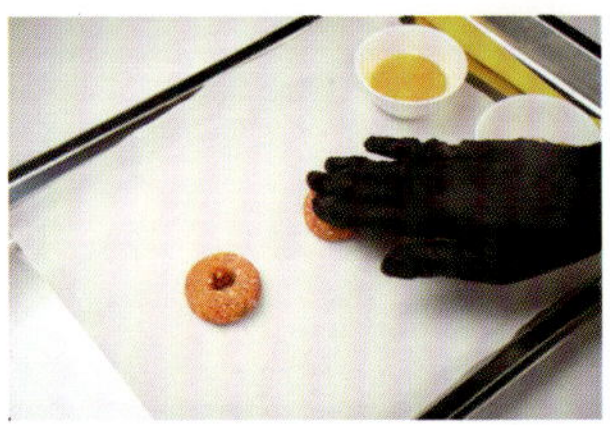

2 取適量紅薯泥整理成圓形，再在中間戳個圓洞，做出甜甜圈的造型。
Separate the sweet potato mixture into small portions. Shape each into a ball, flatten slightly and make a hole in the center to form a donut shape.

3 在甜甜圈上刷上蛋黃液。
Brush donut dough with beaten egg yolk.

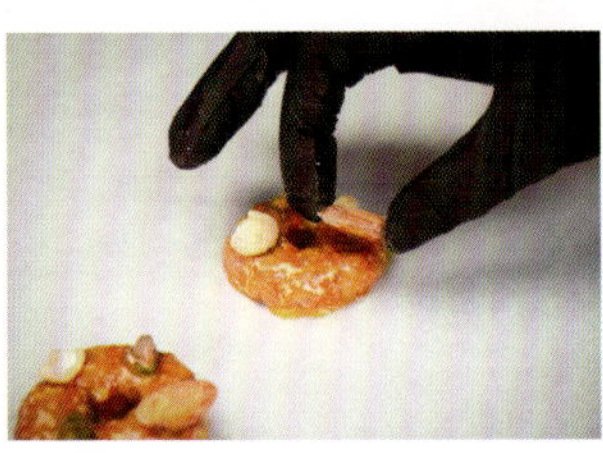

4 用堅果裝飾甜甜圈表面。
Garnish the donuts with mixed nuts.

5 烤箱預熱，上下火170°C烤16分鐘。
Pre-heat the oven, and bake at 170°C for 16 minutes.

優點及營養成份

紅薯本身有足夠的甜味，可以代替糖分來滿足對食品口感的需求。同時含有豐富的維他命A、鉀等營養元素，搭配燕麥和堅果增加了膳食纖維和優質蛋白，有助維持健康。

抹茶燕麥餅乾

間餐系列 Snacks

食材份量 / Ingredient

黑巧克力 5g/克
Dark Chocolate

雞蛋 1pc/隻
Egg

抹茶粉 5g/克
Matcha Powder

桂格即食燕麥片 30g/克
Instant Oatmeal

蜂蜜 5g/克
Honey

Matcha Oat Cookies
抹茶燕麥餅乾

製作步驟 / Step

1 碗中放入燕麥片，打入一個雞蛋，放入蜂蜜；攪拌均勻。
Mix the oatmeal, egg, and honey in a bowl until well combined.

2 將黑巧克力切碎，放入，同時加入抹茶粉，攪勻。
Add the chopped dark chocolate and matcha powder, then mix it well.

3 將抹茶燕麥餅放入烤盤中整理成餅乾形狀。
Shape the dough into cookie shape.

4 烤箱預熱後放入，180°C烘烤15分鐘；翻面再次180°C烘烤15分鐘；取出即可。
Pre-heat the oven. Bake at 180°C for 15 minutes, then flip the cookies and bake for another 15 minutes at 180°C.

優點及營養成份

抹茶燕麥餅乾以燕麥代替麵粉，且雞蛋中含有豐富的蛋白質，黑巧克力中含有豐富的天然抗氧化劑多酚，幫助控制飲食中的脂肪及熱量攝取，有助維持健康。抹茶與巧克力的搭配，符合多數年輕人對美食的喜愛。

蔓越莓燕麥鬆餅

間餐系列 Snacks

食材份量 / Ingredient

牛油 12g/克
Butter

低筋麵粉 40g/克
Low Gluten Flour

泡打粉 5g/克
Baking Powder

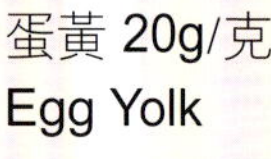

蛋黃 20g/克
Egg Yolk

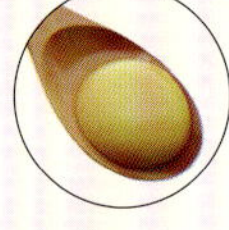

蔓越莓 10g/克
Cranberry

牛奶 30ml/毫升
Milk

桂格即食燕麥片 20g/克
Instant Oatmeal

雞蛋 1pc/隻
Egg

鹽 1g/克
Salt

砂糖 4g/克
Sugar

Cranberry Oat Scones
蔓越莓燕麥鬆餅

製作步驟 / Step

1 將所有原料拌勻，搓成團壓平，切成三角形。表面刷蛋黃液。
Mix all the ingredients then flatten the dough. Cut into triangle shape and brush the surface with beaten egg yolk.

2 預熱烤箱至180°C-200°C，烘烤25分鐘。
Pre-heat the oven, bake at 180°C-200°C for 25 minutes.

優點及營養成份

這款蔓越莓燕麥鬆餅表皮酥脆，內部綿軟，品嘗起來既有燕麥的穀物香氣，又能感受到奶香四溢。蔓越莓的點綴不僅增添了自然的甜味，還增加了抗氧化劑和膳食纖維。這個食譜注重低糖分和適度脂肪，使其成為一種既美味又健康的下午茶選擇。

楊枝甘露燕麥撻

食材份量 / Ingredient

餡料食材 / Fillings

15g/克 - 1:4泡發的桂格即食燕麥片
Instant Oatmeal

雞蛋
Egg 1pc/隻

芒果果茸
Mango 45g/克

淡奶油 60g/克
Light Cream

牛奶 30ml/毫升
Milk

砂糖 15g/克
Sugar

撻底食材 / Tart Base

桂格即食燕麥片 20g/克
Instant Oatmeal

蛋糕粉 120g/克
Cake Flour

雞蛋
Egg 1pc/隻

植物油 30ml/毫升
Vegetable Oil

砂糖 30g/克
Sugar

裝飾食材 / Food Garnish

芒果粒 30g/克
Mango

柚子粒 15g/克
Grapefruit

西米 15g/克
Sago

Mango, Pomelo, and Sago Oat Tart
楊枝甘露燕麥撻

製作步驟 / Step

1 **餡料**：把所有原料拌勻待用。
撻底：所有原料拌勻成團，按壓成0.5cm厚的面皮，放入撻殼模具。
Filling : Mix all the ingredients for later use.
Tart shell : Mix all the tart base ingredients into dough, flatten the dough to 0.5cm and place in the tart mold.

2 將撻底放入烤箱，160°C烤8-10分鐘。冷卻後放入餡料，160°C烤15-20分鐘，表面蛋液凝結不晃動即可。
Bake the tart shell at 160°C for 8-10 minutes. Add the filling after cooling down the tart shell then bake at 160°C for 15-20 minutes until the filling is set.

3 出爐後放上西米、柚子粒及芒果粒裝飾。
Garnish the tart with sago, grapefruit grains and mango before serving.

優點及營養成份

楊枝甘露燕麥撻融合了多種食材，既迎合了口感的需求，又帶來了豐富的營養益處。芒果和柚子為食物提供了天然的酸甜味，同時豐富了維他命和抗氧化物質。此外，燕麥和蛋糕粉為食物提供了多樣性的碳水化合物，而牛奶和淡奶油則為其增添了豐富的蛋白質和鈣質。

豆乳桂花燕麥撻

食材份量 / Ingredient

蛋撻皮 / Tart Sell

桂格即食燕麥片 160g/克
Instant Oatmeal

低筋麵粉 30g/克
Low Gluten Flour

低脂牛奶 20ml/毫升
Low Fat Milk

雞蛋 1/2pc/隻
Egg

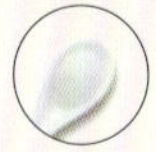
砂糖 3g/克
Sugar

蛋撻液 / Egg Liquid

雞蛋 1/2pc/隻
Egg

低脂牛奶 20ml/毫升
Low Fate Milk

豆漿 10ml/毫升
Soya Milk

砂糖 3g/克
Sugar

香草精 0.1g/克
Vanillon

豆乳 / Soya Paste

豆漿 60ml/毫升
Soya Milk

桂花 1g/克
Osmanthus

砂糖 3g/克
Sugar

玉米澱粉 2g/克
Corn Flour

Soya Milk and Osmanthus Oat Tart
豆乳桂花燕麥撻

製作步驟 / Step

1 **蛋撻皮**：①打勻蛋液後加入燕麥、低筋麵粉、低脂牛奶和砂糖攪拌均勻，靜置5分鐘，分成6等份放入紙杯中整理成杯狀。②預熱烤箱，上下火180°C烤10分鐘定型，晾涼後脫模備用。

Tart Shell : ①Mix the oatmeal, low gluten flour, low fat milk and sugar into beaten egg and rest for 5 minutes. Divide into 6 portions. ②Pre-heat the oven and bake at 180°C for 10 minutes, remove from the mold after cooling down for later use.

2 **蛋撻液**：①所有食攪拌均勻，過篩後將蛋撻液倒入撻皮中。②烤箱預熱，上下火180°C烤20分鐘出爐，晾涼備用。

Egg Liquid : ①Mix all the ingredients, fill each tart shell to 80% full. ②Pre-heat the oven and bake at 180°C for 20 minutes, cool it down for later use.

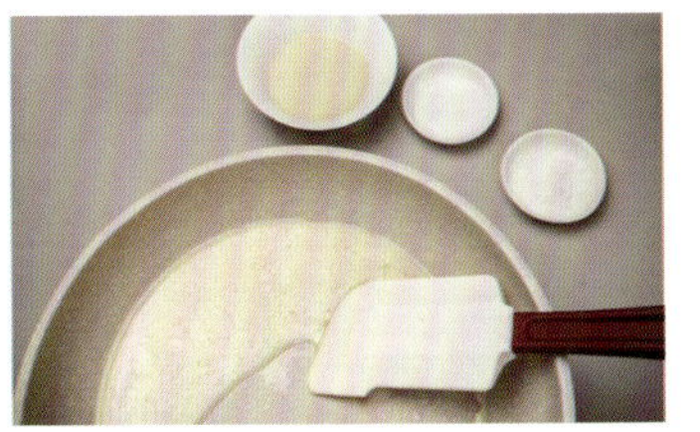

3 **豆乳**：豆漿、砂糖和玉米澱粉攪拌均勻後倒入鍋中，小火加熱攪拌至濃稠。

Soya Paste : Mix soya milk, sugar and corn flour. Heat the mixture over low heat until thickened.

4 最後，在蛋撻表面淋上豆乳，撒少許桂花即可。

Pour the soya paste over the tart and sprinkle with osmanthus before seving.

優點及營養成份

選用燕麥代替蛋撻皮中的麵粉，除幫助調整食物中 GI，還能增加飽腹感；低脂牛奶和豆漿代替黃油、淡奶油等高脂肪食材，降低了脂肪尤其是飽和脂肪酸攝入量；豆漿為植物性優質蛋白。豆乳的香氣中加上一絲絲桂花清香，是下午茶的不錯選擇。

抹茶燕麥糯米糍

Snacks 間餐系列

食材份量 / Ingredient

糯米皮 1pc/片
Glutinous Rice Skin

淡忌廉 4g/克
Whipping Cream

桂格即食燕麥片 8g/克
Instant Oatmeal

抹茶粉 0.5g/克
Matcha Powder

砂糖 1.5g/克
Sugar

温水 24ml/毫升
Warm Water

Matcha Oatmeal Mochi
抹茶燕麥糯米糍

製作步驟 / Step

1 燕麥片用24毫升温水泡發，加入糖拌勻。
Soak the oatmeal in 24ml warm water, then stir in the sugar.

2 加入打發好的淡忌廉，拌勻。
Mix in the whipping cream until well combined.

3 在糯米皮中包入燕麥餡料。
Place the filling onto the glutinous rice skin.

4 撒上抹茶粉即可。
Sprinkle with matcha powder before serving.

優點及營養成份

抹茶燕麥糯米糍在燕麥片中加入了抹茶粉，味道清新，茶香濃郁，略帶苦澀。即食燕麥片的口感與抹茶粉的細膩口感相融合，令人耳目一新，燕麥片不僅能夠增加飽腹感，而且可幫助控制熱量攝取，對於有飲食控制需求的人非常友好，可以作為小甜品的替代。

肉桂蘋果燕麥飲

食材份量 / **Ingredient**

肉桂粉 2g/克
Cinnamon Powder

蘋果果茸 60g/克
Apple Puree

低脂牛奶 200ml/毫升
Low Fat Milk

桂格即食燕麥片 25g/克
Instant Oatmeal

花生醬 5g/克
Peanut Butter

Cinnamon Apple Oat Drink
肉桂蘋果燕麥飲

製作步驟 / Step

1 150毫升低脂牛奶和燕麥片混合後，微波爐加熱1分鐘。
Mix the oatmeal with 150ml low fat milk, heat in microwave for 1 minute.

2 50毫升牛奶、蘋果果茸、花生醬、肉桂粉及泡好的牛奶燕麥片放入攪拌機打均勻便可。
Add 50ml low fat milk, apple puree, peanut butter, and cinnamon powder to the oatmeal and milk mixture. Blend in a mixer until smooth.

優點及營養成份

肉桂蘋果燕麥飲中選用了低脂牛奶來代替普通牛奶，在傳統飲品的基礎上進行了創新。通過添加蘋果和肉桂，增添了新的口感和香氣，飲品中除了燕麥片所能帶來的飽腹感之外，牛奶和花生醬的組合使飲品中鈣含量豐富。

桂花栗蓉燕麥飲

間餐系列
Snacks

食材份量 / **Ingredient**

糖桂花 10g/克
Osmanthus Sugar

錫蘭紅茶 200ml/毫升
Ceylon Tea

燕麥片 20g/克
Oatmeal

桂花 2g/克
Osmanthus

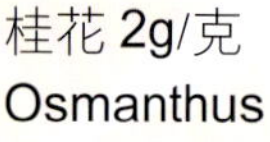

蜂蜜 10g/克
Honey

栗子 60g/克
Chestnut

水 40ml/毫升
Water

牛奶 (打奶泡)
Milk 30ml/毫升

1:4泡發的桂格即食燕麥片 20g/克
Instant Oatmeal

牛奶 100ml/毫升
Milk

Osmanthus Chestnut Oat Drink
桂花栗蓉燕麥飲

製作步驟 / Step

1 板栗提前蒸熟後，加糖桂花、泡發的燕麥片和水打成泥入杯。
Put the osmanthus sugar, cooked chestnut, soaked oatmeal and water in a mug.

2 熱牛奶、錫蘭紅茶、燕麥片和蜂蜜攪勻後倒入杯中。
Add hot milk, Ceylon tea, oatmeal, and honey into the mug.

3 放入奶泡，板栗切碎作為裝飾，撒上桂花。
Top with foam and garnish with chopped chestnuts and osmanthus before serving.

優點及營養成份

這款飲品的食材豐富，融合了燕麥的穀香、錫蘭紅茶的茶香、栗蓉細膩的口感和桂花淡淡的清香。飲品製作過程比較簡單，非常營養健康。

燕麥南瓜甘露

食材份量 / Ingredient

間餐系列 Snacks

椰子水 200ml/毫升
Coconut Water

西米 25g/克
Sago

椰漿 80ml/毫升
Coconut Cream

蜂蜜 5g/克
Honey

南瓜泥 50g/克
Pumpkin Puree

桂格即食燕麥片 35g/克
Instant Oatmeal

Pumpkin Oat Drink
燕麥南瓜甘露

製作步驟 / Step

1 西米煮好加南瓜泥放入杯中。
Add the cooked sago into pumpkin puree and mix well.

2 椰漿、椰子水混勻加熱後加入燕麥片泡發5分鐘，加入蜂蜜，倒入杯中。
Use hot coconut cream and coconut water to soak the oatmeal for 5 minutes. Add honey and mix well.

優點及營養成份

燕麥片和南瓜富含膳食纖維，幫助促進消化和腸道健康。這款飲品具有濃郁的椰香，口感綿密，能夠增加飽腹感，有助於有體重管理需求的人。且其製作方法簡單易行，只需要將燕麥、南瓜和椰子水等混合在一起即可。是一個營養健康兼顧的飲品選擇。

午晚餐系列
Main Dishes

燕麥牛油果沙律

午晚餐系列

Main Dishes

食材份量 / Ingredient

芒果 20g/克
Mango

大米 44g/克
Rice

蝦仁 4pcs/隻
Prawns

柚子粒 30g/克
Grapefruit

腰果 10g/克
Cashew

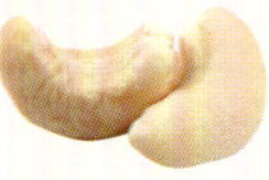

牛油果 50g/克
Avocado

芥末醬 適量
Dijon Mustard a pin

水 75ml/毫升
Water

藜麥 10g/克
Quinoa

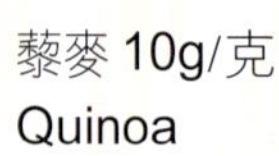

桂格即食燕麥片 12g/克
Instant Oatmeal

Avocado Oat Salad
燕麥牛油果沙律

製作步驟 / **Step**

1 將燕麥片、大米和水放入電飯煲快煮40分鐘，冷卻待用。
Heat the oatmeal, rice with water in rice cooker for 40 minutes. Let cool for later use.

2 將蝦仁、牛油果、腰果、煮熟的藜麥、芒果、西柚、法式芥末醬放在燕麥米飯中，稍加拌勻即。
Add the prawns, avocado, cashews, cooked quinoa, mango, grapefruit, Dijon Mustard to the oatmeal rice. Mix well before serving.

優點及營養成份

牛油果含有多不飽和脂肪酸，大蝦是優質蛋白的來源，腰果富含多種礦物質。一份燕麥牛油果沙律營養豐富均衡，口感清新，果香撲鼻。對於喜歡嘗試新口味和注重健康飲食的人，不妨可以嘗試動手製作。

燕麥香菌菇雞肉三文治

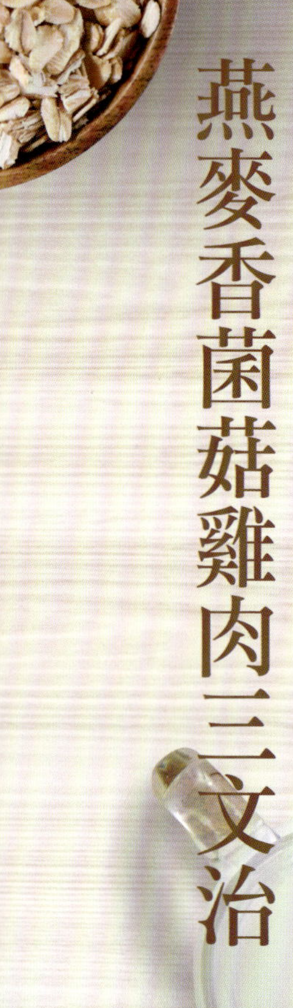

午晚餐系列
Main Dishes

食材份量 / Ingredient

香菇 1pc/隻
Dried Mushroom

雞胸肉 40g/克
Chicken Breast

切片吐司 1pc/片
Sliced Toast

雞蛋 1pc/隻
Egg

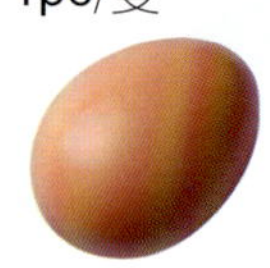

胡椒粉 0.5g/克
Pepper

鹽 1g/克
Salt

葵花籽油 1.2ml/毫升
Sunflower Oil

桂格即食燕麥片 40g/克
Instant Oatmeal

牛奶 150ml/毫升
Milk

Mushroom and Chicken Oat Sandwich
燕麥香菌菇雞肉三文治

製作步驟 / Step

1 燕麥片用熱牛奶浸泡漲發後加入雞蛋，拌勻後，加入鹽、胡椒粉調味。
Soak the oat milk in hot milk, then mix with the beaten egg. Add salt and pepper to increase favor.

2 吐司略煎，雙面微黃即可。
Pan fried the toast till both sides are slightly golden.

3 鍋中放入葵花籽油，放入香菇、雞胸肉炒香，然後放入燕麥片、牛奶、蛋液一起翻炒，炒熟後出鍋。
Spray sunflower oil over pan, stir dried mushroom and chicken breast until cooked, then add the oatmeal with milk, and beaten egg.

4 將炒好的燕麥、香菇及雞胸肉放在吐司上即完成。
Place the cooked mushroom, chicken breast, and oatmeal mixture on the toast before serving.

優點及營養成份

燕麥菌菇雞肉三文治的食材選擇非常豐富。雞蛋、雞胸肉等食材含有豐富的優質蛋白，牛奶中鈣含量十分豐富。此外，即食燕麥片還含有多種維他命。午餐一份三文治果腹，美味、營養兩者兼得！

燕麥蘆筍鮮蝦羹

食材份量 / Ingredient

蘆筍 400g/克
Asparagus

蝦仁 4pcs/隻
Prawns

黑胡椒 1g/克
Black Pepper

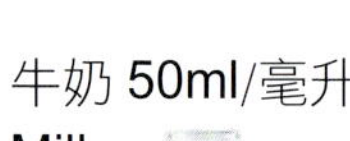

牛奶 50ml/毫升
Milk

雞湯 300ml/毫升
Chicken Soup

桂格即食燕麥片 35g/克
Instant Oatmeal

Asparagus and Shrimp Oat Soup
燕麥蘆筍鮮蝦羹

製作步驟 / Step

1 蘆筍300克用攪拌機打碎。100克切片備用。
Blend 300g and slice 100g of asparagus for later use.

2 大蝦去掉頭和殼，切段炒熟。
Remove the heads and shells from the prawns, cut into pieces and set aside.

3 加入300毫升雞湯和燕麥片，煮開後關火，焗10分鐘至燕片發漲。
Add 300ml chicken soup and oatmeal, bring to a boil and rest for 10 minutes.

4 加入300克蘆筍蓉，慢慢煮開。
Add 300g asparagus puree, stri well and cook gently.

5 加入牛奶，蘆筍片，放入鹽、黑胡椒調味。
Add the milk, reserved asparagus, salt and black pepper before serving.

優點及營養成份

燕麥蘆筍蔬菜羹味道鮮美，大蝦和牛奶富含蛋白質，燕麥的穀香與蘆筍的清爽口感完美融合，加上大蝦的豐富味道，讓人食慾大增。這道菜式可提供優質蛋白質及膳食纖維，簡單易做，是追求健康的人不錯的選擇。

豌豆泥小蘑菇沙律

食材份量 / Ingredient

豌豆 50g/克
Peas

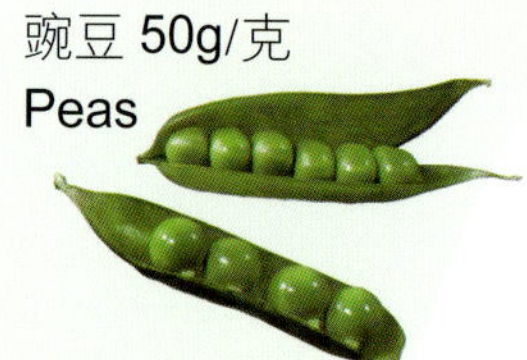

白玉菇 10g/克
White Shimeji Mushroom

雞蛋 1pc/隻
Egg

鹽 1g/克
Salt

生菜 30g/克
Lettuce

黑胡椒 0.5g/克
Black Pepper

橄欖油 5ml/毫升
Olive Oil

桂格即食燕麥片 80g/克
Instant Oatmeal

水 350ml/毫升
Water

Mashed Peas and Mushroom Salad
豌豆泥小蘑菇沙律

製作步驟 / Step

1 豌豆煮熟，將煮熟的豌豆、燕麥片加水放入攪拌機攪打均勻。
Blend the oatmeal, cooked peas, and water in mixer.

2 雞蛋下開水煮熟，切半，白玉菇煎熟備用，生菜切絲；將豌豆泥裝盤，撒上鹽、黑胡椒調味，然後將白玉菇、雞蛋、生菜碼在豌豆泥上即可。
Boil the egg and cut in half. Fry the white shimeji mushrooms and chop the lettuce for later use. Spread the mashed pea puree on a plate and season with salt and black pepper. Add the white shimeji mushrooms, egg, and lettuce on the mashed peas puree before serving.

優點及營養成份

豌豆泥小蘑菇沙律中的食材營養均衡，具有較高的飽腹感，有助於控制飲食能量。豌豆泥的細膩口感和小蘑菇的鮮美口感相互搭配，使得整個沙律更加美味可口。

三文魚燕麥炒飯

午晚餐系列
Main Dishes

食材份量 / Ingredient

杏鮑菇 30g/克
King Oyster Mushroom

胡蘿蔔 20g/克
Carrot

青菜 1pc/顆
Vegetable

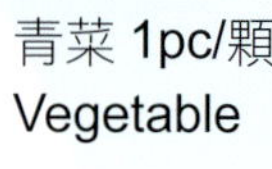

鹽 1g/克
Salt

三文魚 180g/克
Salmon

桂格即食燕麥片 30g/克
Instant Oatmeal

黑胡椒 2g/克
Black Pepper

Fried Rice With Oats and Salmon
三文魚燕麥炒飯

製作步驟 / Step

1 將三文魚、胡蘿蔔、杏鮑菇切粒；青菜切段備用。
Chop the salmon, carrot, king oyster mushroom, and vegetables into piece for later use.

2 三文魚先用平底鍋煎熟，切碎備用。
Fry the chopped salmon in a pan until cooked.

3 青菜、燕麥片和三文魚一起翻炒，炒熟後加鹽、黑胡椒調味。
Stir fried salmon, oatmeal, and vegetables. Season with salt and black pepper before serving.

優點及營養成份

燕麥炒飯中的三文魚除了含有豐富的蛋白質之外，也富含 EPA 和 DHA 等不飽和脂肪酸，利於控制膽固醇的攝入。鮮香魚肉可以使得炒飯口感獨特，營養美味。

豆腐燕麥茄汁燜飯

午晚餐系列
Main Dishes

食材份量 / Ingredient

雞蛋 1pc/隻
Egg

豆腐乾 1pc/件
Dried Tofu

香菜（芫荽）
Coriander 2g/克

蔥 2g/克
Green Onion

玉米粒 40g/克
Corn Kernels

番茄 1pc/隻
Tomato

脆肉瓜 40g
Crispy Meat Melon

桂格即食燕麥片 105g
Instant Oatmeal

調味料 / Seasoning

蠔油 0.5g/克
Oyster Sauce

酒 1ml/毫升
Cooking Win

生抽 1ml/毫Ⅎ
Soy Sauce

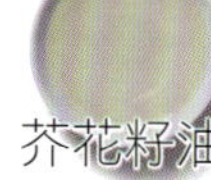
芥花籽油 8g/克
Rapeseed Oil

水 210ml/毫升
Water

Tofu and Tomato Braised Rice With Oats
豆腐燕麥茄汁焖飯

製作步驟 / Step

1 雞蛋煎蛋成餅與豆腐乾、脆肉瓜切粒備用。

Fry the beaten egg and cut into round shape, chop the dried tofu and crispy meat melon for later use.

2 電飯煲中塗抹菜籽油，放燕麥片、水、豆腐乾及所有蔬菜，倒入料酒，啟動電飯煲蒸飯功能。

Spray the rapeseed oil over rice cooker. Add the oatmeal, water, chopped dried tofu, corn kernels, tomato, crispy meat melon, and cooking wine. Start the rice steaming function.

3 待燕麥飯蒸好後，放入生抽、蠔油調味，上面撒蔥花、香菜即可。

After cooking, season with soy sauce and oyster sauce. Sprinkle with green onion and coriander before serving.

優點及營養成份

燕麥片不僅可以用來煮粥，蒸成燕麥飯口感同樣很好。含有更多β-葡聚醣和膳食纖維，升糖指數遠低於白米飯。豆腐乾和雞蛋提供了豐富的蛋白質。番茄和脆肉瓜作為蔬菜，豐富了焖飯的營養素，番茄為這道菜提供了酸甜口感。

奶香菇燕麥燴飯

午晚餐系列
Main Dishes

食材份量 / Ingredient

玉米粒 30g/克
Corn Kernels

西蘭花 50g/克
Broccoli

黑胡椒 1g/克
Black Pepper

香菇 4pcs/隻
Dried Mushroom

鹽 0.5g/克
Salt

混合海鮮（蝦仁、魷魚） 100g/克
Mix Seafood (Shrimp / Squid)

生抽 1ml/毫升
Soy Sauce

低脂牛奶
Milk 120ml/毫升

桂格即食燕麥片 60g/克
Instant Oatmeal

水 50ml/毫升
Water

Creamy Mushroom Oat Risotto
奶香菇燕麥燴飯

製作步驟 / **Step**

1 香菇、西蘭花切粒備用。
Chop the mushrooms and broccoli into small pieces for later use.

2 鍋中倒入適量油，放入香菇和海鮮翻炒。
Spray oil over pan, stir-fry the mushrooms and seafood.

3 加入120毫升低脂牛奶和50毫升水，待水開後加入玉米粒、西蘭花煮熟。
Boil the 120ml low-fat milk with 50ml water, and add the corn kernels and broccoli and cook until tender.

4 關火，放入燕麥片、生抽、鹽和黑胡椒攪拌均勻，蓋上鍋蓋焗3分鐘即可享用。
Turn off the heat. Add the oatmeal, soy sauce, salt and black pepper and mix well. Cover the lid and let sit for 3 minutes before serving.

優點及營養成份

低脂奶油代替傳統奶油，經過熬製的牛奶濃稠度增加，奶香十足，並且蛋白質及鈣含量十分豐富。混合小海鮮是優質蛋白的來源，香菇、西蘭花等蔬菜的加入使整個燴飯營養均衡全面。最後加入燕麥，避免了長時間高溫加熱導致燕麥升糖指數提高的問題。

燕麥福鼎肉片

Main Dishes 午晚餐系列

食材份量 / Ingredient

裙帶菜 1g/克
Wakame

木薯粉 50g/克
Tapioca Flour

原隻大蝦
Prawns 1pc/隻

生抽 1ml/毫升
Soy Sauce

青菜 100g/克
Vegetable

麻油 1ml/毫升
Sesame Oil

鹽 0.5g/克
Salt

雞胸肉 100g/克
Chicken Breast

桂格即食燕麥片 50g/克
Instant Oatmeal

Fuding Chicken and Oat Slices Soup
燕麥福鼎肉片

製作步驟 / Step

1 燕麥片、木薯粉、雞胸肉、生抽一起用絞肉機攪打上勁，倒入適量冰水多次攪打備用。
Mix the oatmeal, tapioca flour, chicken breast, and soy sauce into a food processor. Gradually add ice water while mixing to from a meat paste.

2 鍋裡燒開水後轉小火，將肉片撥入鍋中，3分鐘肉片浮起即可盛盤備用。
Boil water, then reduce to low heat, add the meat slices, and cook for 3 minutes until they float.

3 在鍋中放入青菜、裙帶菜、大蝦、鹽和麻油，加開水煮沸。
Add the vegetables, wakame, prawns, salt and seasame oil to the pot and cook until done.

4 放入肉片即可食用。
Add the cooked meat slice before serving.

優點及營養成份

這是一道改良版的燕麥福鼎肉片，肉片採用脂肪較低的雞胸肉。肉片裡添加了燕麥，比普通的福鼎肉片含有更多的膳食纖維，同時保證了一餐主食的需求。與普通的福鼎肉片相比，本菜式增加了深色蔬菜，更加滿足一餐蔬菜攝入量以及營養素的需求。

馬蹄燕麥蒸肉丸

食材份量 / Ingredient

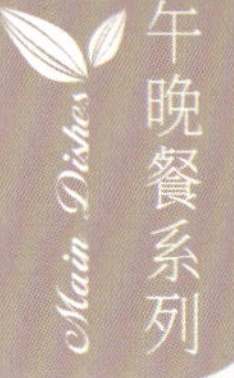

馬蹄 15g/克
Chinese Water Chestnut

澱粉 3g/克
Starch

胡椒粉 0.5g/克
Pepper

豬肉 50g/克
Pork

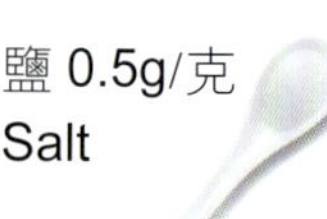

鹽 0.5g/克
Salt

芋頭
Taro 50g/克

桂格即食燕麥片 25g/克
Instant Oatmeal

水 25ml/毫升
Water

Steamed Water Chestnut Oat Meatballs
馬蹄燕麥蒸肉丸

製作步驟 / Step

1 豬肉餡放鹽、胡椒粉調味，加25毫升水攪拌至黏。
Season pork with salt and pepper, then add 25ml water and mix until the mixture becomes sticky.

2 馬蹄去皮，拍碎成小顆粒狀；芋頭去皮切成厚粒。
Remove the skin of Chinese water chestnut then chop into small pieces. Remove the skin of taro then chop into larger pieces.

3 肉餡中放入澱粉、馬蹄碎、芋頭碎，攪拌均勻。
Add starch, chopped Chinese water chestnut, and taro into the meat mixture.

4 肉餡擠成核桃大的小丸子，沾上燕麥片。
Shape the mixture into chestnut-sized meatball and cover with the oatmeal.

5 約蒸20分鐘即可。
Steam for 20 minutes and ready to serve.

優點及營養成份

燕麥與豬肉搭配具有營養價值互補。肉丸搭配脆嫩的小馬蹄和軟糯的芋頭，口感豐富，營養全面。蒸的烹飪方式能夠保留食材的原汁原味，使得肉丸更加鮮美可口，能夠為人們提供一種全新的健康飲食的選擇。

燕麥乾煎三文魚

午晚餐系列 *Main Dishes*

食材份量 / Ingredient

三文魚 100g/克
Salmon

雞蛋 1pc/隻
Egg

葵花籽油 3ml/毫升
Sunflower Oil

桂格即食燕麥片 30g/克
Instant Oatmeal

鹽 0.5g/克
Salt

黑胡椒 0.5g/克
Black Pepper

Pan-Seared Oat-Crusted Salmon
燕麥乾煎三文魚

製作步驟 / Step

1 沖洗三文魚，用廚房紙吸去表面水份，切成塊。
Wash salmon under running water and dry it with a kitchen towel before chopping into portions.

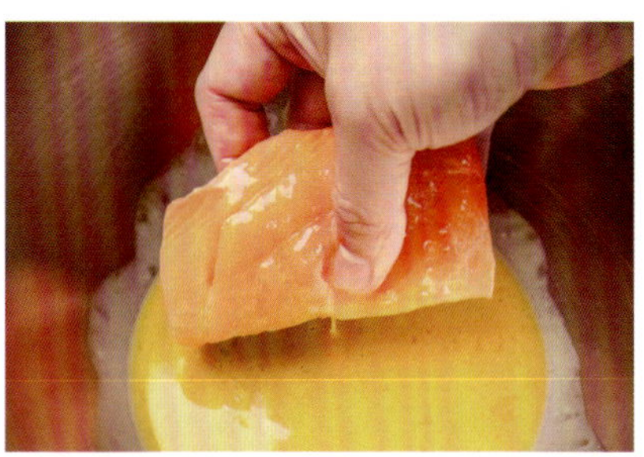

2 雞蛋打散，三文魚塊沾上蛋液。
Brush beaten egg over salmon.

3 將沾滿蛋液的三文魚塊繼續沾滿燕麥片。
Cover the salmon with oatmeal.

4 平鍋預熱，倒入少許油，放入沾好燕麥片的三文魚塊，小火慢煎。將每個表面煎至金黃即可。放入鹽、黑胡椒調味。
Spray oil over pre-heat pan, place the oatmeal salmon and cook over low heat until both sides are golden brown. Season with salt and black pepper before serving.

優點及營養成份

三文魚是非常優質的一種食材，富含多種營養成分，尤其富含 Omega-3 脂肪酸，尤其是 EPA 和 DHA。

燕麥西蘭花
三文魚忌廉湯

食材份量 / Ingredient

午晚餐系列
Main Dishes

洋蔥 1/2 pc/隻
Onion

西蘭花 300g/克
Broccoli

三文魚 250g/克
Salmon

桂格快熟燕麥片 30g/克
Quick Cook Oatmeal

雞湯 1L/公升
Chicken Stock

淡忌廉 100g/克
Whipping Cream

調味料 / Seasoning

黑胡椒 適量
Black Pepper a pinch

鹽 適量
Salt a pinch

Salmon & Broccoli Oat Cream Soup
燕麥西蘭花三文魚忌廉湯

製作步驟 / Step

1 西蘭花及洋蔥洗淨，西蘭花切成細朵、洋蔥切粒，備用。
Clean and chop broccoli and onion into pieces for later use.

2 鍋中下油燒熱，加入洋蔥炒香，至洋蔥至透明度狀。倒入雞湯煮滾，放入西蘭花，再慢火煮約 5-6 分鐘至西蘭花軟身。
Spray oil over pan, stir-fried onion into transparent form. Add chicken stock and bring to a boil, then add the broccoil. Reduce to low heat and cook 5-6 minutes until softened.

3 煮湯期間，將三文魚放入預熱 200 度的焗爐，焗約 6-8 分鐘至熟，用叉將魚肉壓爛，備用。
Pre-heat the oven to 200°C. Bake the salmon for 6-8 minutes, then flake with a fork.

4 將湯用手提攪拌機或倒入攪拌機中，打至材料完全幼滑。
Blend the soup mixture (broccoli, onion, and chicken stock) in a blender until smooth.

5 加入燕麥、三文魚鬆及淡忌廉以慢火煮滾至自己喜愛的濃稠度，最後加入適量鹽及黑胡椒調味，試味可以後，即可盛起享用。
Add the cooked oatmeal, flaked salmon, and whipping cream. Simmer over low heat until the desired taste is reached. Season with salt and black pepper before serving.

小貼士 / Tips

此湯除了選用西蘭花之外，還可以選用椰菜花、粟米、南瓜、薯仔等其他自己喜愛的蔬菜代替，做法大致一樣，但各種蔬菜做出來的濃稠度有別，所以份量可按個人喜好調校。

養生五色燕麥菜飯

食材份量 / Ingredient

午晚餐系列 *Main Dishes*

小棠菜 100g/克
Baby Pak Choi

淮山粒 50g/克
Diced Chinese Yam

水 400ml/ 毫升
Water

玉米粒 60g/克
Corn Kernels

甘筍粒 40g/克
Diced Carrot

雞髀肉（切粒）100g/克
Diced Chicken Thigh

桂格燕麥飯 100g/克
Oats For Rice

米 200g/克
Rice

羊肚菌 10g/克
#Morel Mushroom

雞髀醃料 Chicken Marinade

胡椒粉 適量
Pepper a pinch

鹽 2g/克
Salt

生抽 4ml/ 毫升
Soy Sauce

砂糖 2g/克
Sugar

浸軟 洗淨切粒 soaked, cleaned and chopped

Five-Colour Oat & Veggie Rice
養生五色燕麥菜飯

製作步驟 / Step

1 小棠菜洗淨，然後放入滾水中煮約 1 分鐘，然後撈起沖水揸乾水份切粒。
Clean and cook baby pak choi in boiling water for 1 minute, then squeeze out the water and dice into pieces.

2 米洗淨與桂格燕麥飯，倒入電飯煲中再加入水及醃好的雞髀肉、五色材料 (甘筍粒、粟米粒、淮山粒、羊肚菌及小棠菜) 拌勻蓋上飯煲蓋，按上煮飯鍵，煲約 50 分鐘，吃時將所有材料與飯撈勻即可趁熱享用。
Wash the rice. Add the rice, oats, marinated diced chicken thigh, 5 color ingerdients (diced carrot, corn kernels, diced Chinese yam, morel mushrooms, and baby pork choi), and water into the rice cooker. Press "cook rice" and cook for about 50 minutes. Stir well before serving.

小貼士 / Tips

1. 小棠菜及其他五色食材煮前均需要揸乾水分，否則煮出來的飯便會太濕不好吃。
2. 五色食材是近年很流行的養生食療，就是均衡吸收分別為紅、黃、綠、白、黑五種顏色的各種食材。

黑胡椒雞粒炒燕麥飯

午晚餐系列

Main Dishes

食材份量 / Ingredient

雞髀肉（切粒）300g/克
Diced Chicken Thigh

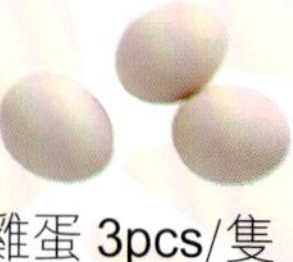

雞蛋 3pcs/隻
Egg

桂格即食燕麥片 100g/克
Instant Oatmeal

醃料 / Marinade

胡椒粉 適量
Pepper a pinch

鹽 2g/克
Salt

生粉 3g/克
Cornstarch

生抽 15ml/ 毫升
Soy Sauce

黑胡椒碎 3g/克
Cracked Pepper

砂糖 3g/克
Sugar

調味料 / Seasoning

鹽 2g/克
Salt

Black Pepper Chicken Fried Oat Rice
黑胡椒雞粒炒燕麥飯

製作步驟 / Step

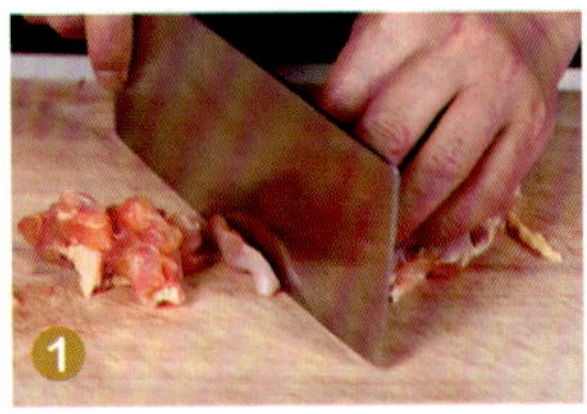

1 雞髀肉洗淨吸乾水份，然後加入醃料拌勻醃製 20 分鐘備用。
Pat the chicken thighs dry after washing, then mix with the marinade for 20 minutes for later use.

2 鑊燒熱加入少許油放入醃好的雞髀肉炒熟撈起。
Spray oil to a heated pan and stir-fry the chicken until cooked through.

3 桂格即食燕麥片與雞蛋及鹽拌勻，然後鍋燒熱加入少許油，倒入拌好的燕麥片炒散後，將雞粒回鍋炒勻即可。
Mix the oatmeal, eggs, and salt. Add a little more oil to the pan, pour in the oat and egg mixture, and stir-fry together with the chicken until everything is cooked and combined.

小貼士 / Tips

除了雞肉外，還可改以各種肉類或海鮮代替，各具風味。

燕麥乾果餅卷

午晚餐系列 *Main Dishes*

食材份量 / Ingredient

水 300ml/ 毫升
Water

食油 適量
Cooking Oil a pinch

砂糖 15g/克
Sugar

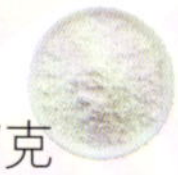

糯米粉 300g/克
Glutinous Rice Flour

粘米粉 或 中筋麵粉 30g/克
Rice Flour or Plain Flour

桂格 5 紅混合即食燕麥片 20g/克
5 Red Multi-Grain

餡料 / Filling

炒香的芝麻 15g/克
Toasted Sesame

炒香的花生 50g/克
Toasted Peanuts

花生醬 適量
Peanut Butter a pinch

草莓乾、杏脯、雜錦提子乾：各適量及切碎
Dried Strawberries, Apricots, and mixed Raisins: a pinch as needed, finely chopped

Oat & Dried Fruit Pancake Rolls
燕麥乾果餅卷

製作步驟 / Step

1 糯米粉、粘米粉 或 中筋麵粉、砂糖及水混合，搓成粉團，再分成適當大小，備用。
Mix glutinous rice flour, rice flour, / plain flour, sugar and water into dough. Then divide into small portions for later use.

2 平底鑊下適量油，不用開火，將粉團放入，再用手或勺將粉團向四面壓平至自己喜歡的厚薄度，然後開中小火，將兩面煎至微焦香脆，成薄卷皮，倒出備用。
Add oil to pan without heating it, place a portion of dough in the pan and flatten the desired thickness, then heat it over medium heat till both sides are crispy.

3 將餅卷皮攤平，略為放涼，先均勻塗上適量花生醬，再均勻放上桂格 5 紅混合即食麥片及其他乾果碎，將薄卷皮捲起成圓狀，然後切段，上碟後即可趁熱享用。
Let the pancakes cool. Spread with peanut butter, sprinkle with 5 red multi-grain oatmeal and dried fruit. Roll up and cut the pancake into pieces before serving.

小貼士 / Tips

1. 要將粉團在開火前放入，否則油已燒熱的話，粉團一放入就開始變熟，便較難造型了。
2. 餅皮更可加入如斑蘭汁，胡蘿蔔汁、綠茶粉、紅菜頭汁等等天然色素，變化出不同顏色及香氣的薄卷。
3. 將薄卷皮捲好之後，用手將薄餅卷輕輕壓一下，這樣會令餡料黏付在薄卷皮上，吃時便不會吃到一地都是餡碎了。

星洲黃金麥皮蝦

午晚餐系列

Main Dishes

食材份量 / Ingredient

大頭蝦 10pcs/隻
Large Prawns

咖喱葉 3g/克
Curry Leaves

雞蛋 1pc/隻
Egg

指天椒（切圈）1pc/隻
Bird's Eye Chilli (sliced)

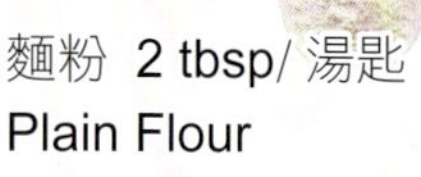

麵粉 2 tbsp/ 湯匙
Plain Flour

桂格 5 紅混合即食燕麥片 100g/克
5 Red Multi-Grain

牛油 50g/克
Butter

鹽 1/2tsp/ 茶匙
Salt

鹹蛋黃 3pcs/隻
#Salted Egg Yolk

蒸 15 分鐘至熟，用叉爛成鹹蛋黃茸
steamed for 15 mins and mashed

調味料 / Seasoning

砂糖 適量
Sugar a pinch

鹽 適量
Salt a pinch

Singapore-Style Golden Oat Prawns
星洲黃金麥皮蝦

製作步驟 / Step

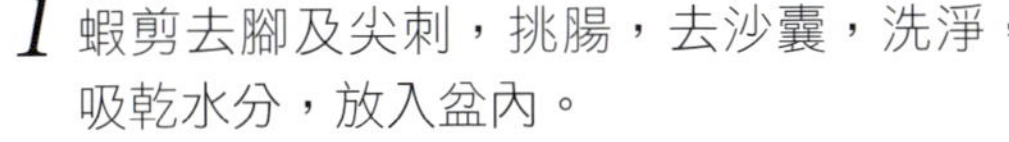

1 蝦剪去腳及尖刺，挑腸，去沙囊，洗淨，吸乾水分，放入盆內。
雞蛋打勻，先與蝦撈勻，再下鹽及逐些加入麵粉再撈勻，令每隻蝦都均勻沾滿雞蛋及麵粉。
Remove prawn legs, spikes, and devein. Wash and pat dry. Mix prawns with beaten egg, then add salt and plain flour, ensuring all prawns are evenly coated.

2 鑊中燒熱適量油，將蝦兩邊煎至約八成熟，撈起，瀝乾油份，備用。
Spray oil over pan and fry the prawns until about 80% cooked for later use.

3 鑊以中小火煮溶牛油，下咖喱葉及指天椒爆香，下鹹蛋黃煮開至起泡，下調味料及桂格 5 紅混合即食麥片抄勻，蝦回鑊，將所有材料與蝦撈勻，令每隻蝦都沾滿麥皮、蛋黃等材料及蝦全熟，即可盛起上碟排好，趁熱享用了。
Melt the butter over medium heat, stir-fry curry leaves and sliced bird's eye chili until fragrant. Add the mashed salted egg yolks, sugar, salt and 5 red multi-grain oatmeal, stir-fry to combine. Put the prawns to the pan and toss until fully coated and cooked through. Serve immediately.

小貼士 / Tips

1. 鹹蛋黃不要壓得太碎，留有少許鹹蛋黃的細碎顆粒口感會更有風味。同時亦要拿走鹹蛋黃中間那粒較硬部分。
2. 咖哩葉在一般賣椰汁、咖喱的香料店買到，若買不到的話，用檸檬葉代替也可以。

燕麥花雕肉餅蒸蟹

食材份量 / Ingredient

午晚餐系列
Main Dishes

免治豬肉 300g/克
Minced Pork

薑米 5g/克
Minced Ginger

雞蛋 1/2pc/隻
Egg

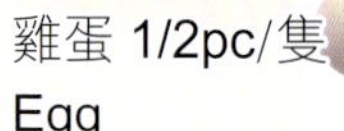

肉蟹 1pc/ 隻 (1kg / 斤)
Mud Crab

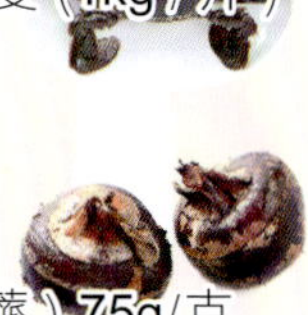

馬蹄（荸薺）75g/克
Chinese Water Chestnut

桂格即食燕麥片 15g/克
Instant Oatmeal

調味料 / Seasoning A

胡椒粉 適量
Pepper a pinch

砂糖 3g/克
Sugar

花雕酒 5g/克
Huadiao Wine(A)

蠔油 5g/克
Oyster Sauce

老抽 1 tsp/ 茶匙
Dark Soy Sauce

水 30ml/ 毫升
Water

糟鹵（酒釀）15g/克
Fermented Rice Wine

調味料 / Seasoning B

花雕酒 8g/克
Huadiao Wine(B)

Steamed Crab with Huadiao Pork & Oats
燕麥花雕肉餅蒸蟹

製作步驟 / Step

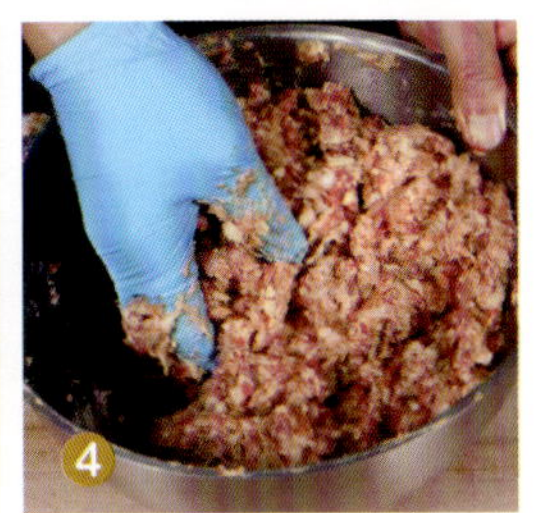

1 蟹劏好洗淨，斬件，瀝乾水分。馬蹄洗淨，去皮，切粒。然後將桂格即食燕麥片、免治豬肉、馬蹄、薑米及雞蛋與調味料 (A)（除水外）攪勻，然後再逐少慢慢加入水，攪至肉餅起膠帶有黏性，再將肉餅「撻」打，令肉餅的質感吃起來更有彈性。

Clean and chop the crab into pieces. Peel and dice the Chinese water chestnuts. In a bowl, mix the oatmeal, minced pork, diced Chinese water chestnuts, minced ginger, egg and all seasoning A ingredients except water, then gradually add water until the mixture becomes sticky and elastic.

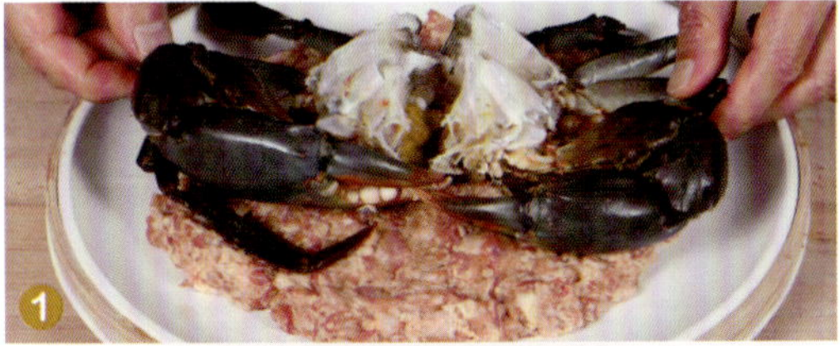
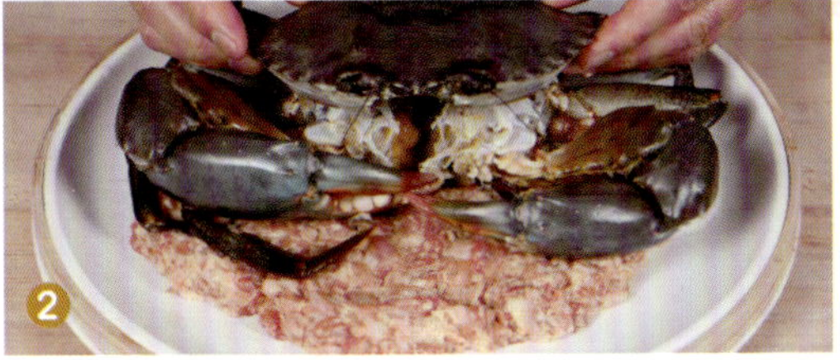

2 將處理好的肉餅鋪在蒸碟上，然後將蟹放在肉餅上，水滾後以大火蒸約 13-15 分鐘（視乎蟹的大小），取出，在肉餅上淋上花雕酒 (B)，最後灑上蔥花即可趁熱享用。

Spread the pork mixture evenly on a plate and arrange crab pieces on top. Steam over high heat for 13-15 minutes (depending on the size of the crab). After steaming, drizzle with huadiao wine (seasoning B) and sprinkle with chopped green onion before serving.

小貼士 / Tips

1. 肉餅蒸出來口感要香滑，關鍵在於豬肉一定要帶有肥肉，同時在調味料上要加入適量桂格即食燕麥片、水分、及雞蛋，這樣肉餅蒸出來便會口感鬆軟及香滑，同時還要將肉餅攪勻及「撻」至起膠狀及帶有黏性，這樣肉餅吃起來才會有彈性。
2. 將蟹放在肉餅上時，蟹肉要輕輕插入肉餅的表面，這樣肉餅便能更吸收到蟹的鮮味了。調味料的花雕酒 (B)，要蒸好後最後才淋上，這樣酒香便不會被揮發，吃時便可聞到陣陣的酒香味了。

養生燕麥・健康飲食指南

作　　者／PepsiCo Beverages Hong Kong Limited
Address: Unit 2912-16, 29/F, Tower A, 83 King Lam Street, Cheung Sha Wan, Kowloon, Hong Kong

編　　譯／JOWIE TO
排　　版／PING MAN
封　　面／阿俊

出　　版／才藝館（匯賢出版）
地址：新界葵涌大連排道144號金豐工業大廈2期14樓L室
Tel : 852-2428 0910
web : https://www.wisdompub.com.hk　email : info@wisdompub.com.hk
search : wisdompub
出版查詢／Tel : 852-9430 6306《Roy HO》

書店發行／一代匯集
地址：九龍旺角塘尾道64號龍駒企業大廈10樓B & D室
Tel : 852-2783 8102　Fax : 852-2396 0050
facebook : 一代匯集　email: gcbookshop@biznetvigator.com

版　　次／2025年7月初版
定　　價／(平裝) HK$128.00　(平裝) NT$630.00
國際書號／ISBN 978-988-75522-6-0
圖書類別／食譜

QUAKER
EST 1877